Radio Frequency Identification and Sensors

To my father

"If you work hard at the small things,
in time you will achieve the great ones".

Samuel Beckett

Series Editor
Guy Pujolle

Radio Frequency Identification and Sensors

From RFID to Chipless RFID

Etienne Perret

ISTE

WILEY

First published 2014 in Great Britain and the United States by ISTE Ltd and John Wiley & Sons, Inc.

ISTE Ltd
27-37 St George's Road
London SW19 4EU
UK

www.iste.co.uk

John Wiley & Sons, Inc.
111 River Street
Hoboken, NJ 07030
USA

www.wiley.com

Library of Congress Control Number: 2014953189

British Library Cataloguing-in-Publication Data
A CIP record for this book is available from the British Library
ISBN 978-1-84821-766-9

Contents

Acknowledgments

This project was completed at the Laboratory of Conception and Integration of Systems (LCIS) at the Grenoble Institute of Technology, in the ORSYS group.

Special thanks go to Hamza Chaabane, Arnaud Vena and Raji Sasidharan Nair, whom I supervised during their PhD degree; the principal results of which are discussed in this book. I extend my warm friendship and gratitude to them.

I would also like to thank all of the master's, doctoral and post-doctoral students with whom I have worked over the past few years and who contributed to the completion of this book. Particular thanks go to Emna Bel Kamel, Maxime Bernier, Ayslan Caisson Noroes Maia, Mossaab Daiki, Thaïs Luana Vidal, Maher Hamdi, Olivier Rance, Taranjeet Singh, Paul Slomianny, Raphael Tavares De Alencar and Deepu Vasudevan Nair. Both the quality and the relevance of this book owe a great deal to their assiduous work, and I am well aware of being in their debt.

Warm thanks to my colleagues Frédéric Garet and Lionel Duvillaret, with whom I have had many fruitful discussions, notably as part of the THID project. I would like to express my appreciation for all their help and for the time they most kindly spared for me. I also give thanks to Thierry Barron, Patrice Gonon and Christophe

Vallée, who also played a role in this work by enabling us to open new paths for chipless RFID.

My sincere thanks also go to the people with whom I have had the opportunity to work on various research contracts, the subjects of which formed the skeleton of this book. I would like to extend my particular thanks to Yann Boutant, Guy Eymin-Petot-Tourtollet, Christophe Halopé, Diez Hubert, Christophe Loussert and Christophe Poyet. Thank you for listening and helping me throughout the years.

I am particularly grateful to my partner Emma, who dedicated herself with patience, diligence and precision to the correction of my manuscript, a task made all the more thankless by not being in her field.

Thanks as well to all my longtime laboratory colleagues and friends, especially the members of the ORSYS group, Darine Kaddour, Pierre Lemaitre Auger, Romain Siragusa and Smail Tedjini, with whom I have spent many hours discussing our respective research. I am very grateful for the highly varied exchanges we have had together through the years, and I extend my deepest friendship and best wishes to them.

This work was supported financially by several research programs, institutions and companies. The principal contributors were the Grenoble Institute of Technology, the French National Research Agency (ANR), Gravit Innovation Grenoble – Alps, the Centre National d'Etudes Spatiales (CNES) and the department of Drôme.

List of Acronyms

ANR	*Agence Nationale de la Recherche* (the French National Research Agency)
CBRAM	Conductive Bridging RAM
CST	computer simulation technology
CTP	*Centre Technique du Papier* (the Pulp and Paper Research & Technical Centre)
EAN 13	European Article Numbering 13 barcode
EAS	Electronic Article Surveillance
EIRP	effective isotropic radiated power
EM	electromagnetic
EPC	Electronic Product Code
ERP	effective radiated power
ETSI	European Telecommunications Standards Institute
FCC	Federal Communications Commission

FCS	Frame Control Sequence
FET	field effect transistor
GD	group delay
GA	genetic algorithms
HF	high frequency
RH	relative humidity
IC	integrated circuit
ID	identifier
IFF	Identification Friend or Foe
ISO	International Organization for Standardization
ISM	industrial scientific and medical
LF	low frequency
LCIS	*Laboratoire de Conception et d'Intégration des Systèmes* (Laboratory of Conception and Integration of Systems)
SEM	scanning electron microscopy
MIM	Metal–Insulator–Metal
MIT	Massachusetts Institute of Technology
MST	Modulated Scattering Technique
NFC	near-field communication
OOK	on–off keying
PE	polyethylene
PET	polyethylene terephthalate

PMC	programmable metallization cell
PIN	positive intrinsic negative diode
PMC	programmable metallization cell
PMMA	polymethylmethacrylate (acrylic)
PPM	pulse-position modulation
PSD	power spectral density
PTFE	polytetrafluoroethylene
RCS	radar cross-section
REP	RF encoding particles
RF	radio frequencies
RFID	radio frequency identification
SAW	surface acoustic wave
SER	surface equivalent radar
SFH	synthesized frequency hopping
SMA	SubMiniature version A
THID	terahertz identification
THz	terahertz
UHF	ultra high frequency
UWB	ultra wide band
VNA	vector network analyzer

Introduction

There is an unprecedented enthusiasm for radio frequency identification (RFID) technologies today. RFID is based on the exchange of information carried by electromagnetic waves between a label, or tag, and a reader. This technology is currently in full economic expansion, which has manifested itself in widely backed research activities, some of which will be examined in this book. There are multiple annual international conferences dedicated specifically to this theme, and RFID sessions are included in every conference revolving around microwaves, RF systems or issues of communication. This can be explained by the versatile quality of this approach, which makes it possible to address extremely broad domains ranging from software to components. Today, there are thousands of applications that involve RFID. Here too, the spectrum is a considerable one, ranging from logistics to passports but also including niche domains, some of which are quite unexpected. This extremely wide variety of applications has yielded a large number of limitations, which differ according to the intended field of use, necessitating the creation of tags of various sizes, able (or not) to resist high mechanical- or temperature-based stresses or to ensure secure data exchange. To respond to these needs, which can sometimes prove incompatible, different RFID technologies have appeared over time. It is why radio frequency technology is pluralist. Simply attempting to categorize RFID technologies into groups is in itself a fairly complicated undertaking. Of these technologies, we will pay particular attention to passive RFID (meaning that no energy source is present on the tag side) and ultra high frequency (UHF) (in which signal exchanges are conducted by propagation and not by coupling). This technology has overtaken HF technology, which has a short reading distance and which makes remote readings more difficult than contact readings.

RFID has been constructed normatively over time, and this standardization is its true strength. Because of this, in the future, RFID will no longer be limited to the domain of identification, but will take on a whole new dimension. Like WiFi, for example, RFID is being regarded widely as a set of wireless communication protocols regulated by norms, over which any type of data can be transmitted. This is only the beginning, and projected applications include the manufacture of sensors with a unique identifier (ID). Among these multiple RFID applications and technologies, however, certain trackers remain that connect things with one another, thus maintaining a certain unity. Cost remains the most important tracer, particularly that of the tag, clearly distinguishing it from other systems such as Bluetooth or ZigBee. In passive UHF RFID, the absence of a battery in the tag results in a much lower cost, and it is possible with many applications to obtain disposable tags. On a note closely related to the cost aspect, it is also true that RFID is intended to constitute a simple approach, notably from an electronic standpoint. This simplicity of production and application might also be considered in a certain way as another RFID tracer.

When we consider the concepts of identification, tag cost and simplicity of application, we are logically led to think of the barcode, which is truly the benchmark reference for this subject, and will serve as the reference point for a large number of discussions in this book. This leads us to the following statement: if we are seeking to push the envelope of the aforementioned RFID codes, is not it relevant for RFID to attempt to move closer to barcodes? Put like that, this may seem like a strange suggestion, but we will see that it is nothing of the kind. The second part of this book will examine another branch of RFID, which we call chipless RFID. We will discuss in detail the reasons that led researchers to work on this unusual issue, which can be situated from an applicative point of view somewhere halfway between RFID and the barcode.

This book is structured into two parts. Part 1 is entirely dedicated to passive UHF RFID. Following an introduction (Chapter 1) to the general approach used in passive RFID, specifically the principle of backscattering, the issue of designing antennas to create tags that are as impervious as possible to the environment will be discussed in Chapter 2. Conversely, the question of tags used to relay information about their environment will be addressed in Chapter 3. Part 2 will be dedicated to chipless technology. After introducing the principle used in Chapter 4, Chapter 5 will examine the various developments that have been applied. The last chapter, Chapter 6, will present the different functionalities toward which it is envisioned that the chipless approach might converge. We will see, for example, how it is possible to produce tag-sensors and reconfigurable tags.

PART 1

Radio-Frequency Identifications

1

Introduction to RFID

1.1. General introduction to RFID

Radio frequency identification (RFID) is a major technology which, for more than a decade now, has undergone significant development in terms of applications. The traceability market includes a large number of families of tags, with each of these families fulfilling very specific needs. These tags include a label composed of an antenna and a medium for information (generally using a silicon chip); some contain a battery (active tag) and some do not (passive tag) [FIN 10, PAR 09, TED 09]. The operating principle is that of a classic wireless communication device. The reader is used to establish an interface between the identifier (ID) contained in the tag on the one side, and, on the other side, a database is used to access other information on the tag. The reader transmits a frame according to a standardized protocol, and the signal is then modulated. In the case of ultra high frequency (UHF) RFID, the tag receives part of this radio-frequency (RF) wave. The tag's antenna then collects the signal and transmits it to the RFID chip (Figure 1.1). For passive RFID tags, the power transmitted by the wave generated by the reader enables the tag to be fed, and thus to function. In addition, the reader transmits the carrier wave continuously, while the frame is sent periodically. In this case, the tag's antenna performs a type of double function: (1) ensuring the tag's power supply and (2) enabling the transmission of the "useful" signal to the chip to be processed.

We will look next at the tags that are most often found on the RFID market, specifically passive tags. Similarly, we will focus mainly on tags that function on UHF frequencies and therefore operate in remote field zones, and can enable communication over distances as small as 10 m. This family of tags is widely deployed, although this deployment has been slowed down mainly by current significant limitations in RFID technology, which will be discussed below. Throughout this book, we will also explore the importance of the tag antenna in an RFID system.

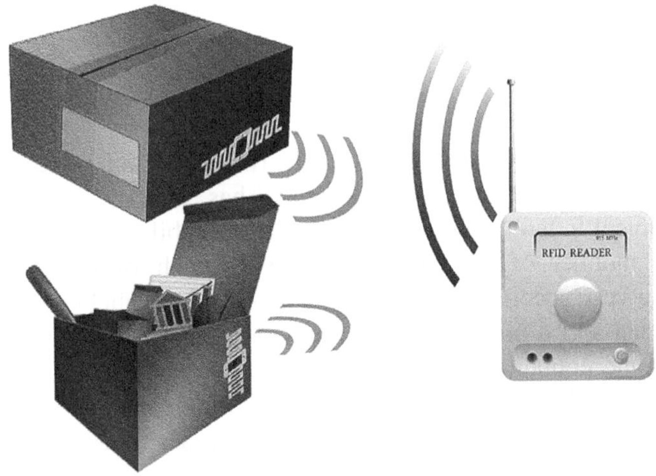

Figure 1.1. *Operation schema of an RFID system. The reader, most often linked to a database, questions the tag, located here on a packing box. In this way, the reader can recover the tag's ID, contained in the RFID chip located in the center of the tag*

Antenna design is at the heart of many of the issues currently being confronted by UHF RFID. This is all the more true, because the design of antennas for UHF RFID tags is different in many ways from classic RF antenna design. Before going into specific details about these differences, note the following two points: (1) the near environment, i.e. the object on which the tag is positioned for its use and which is unknown to the designer, although the tag must be able to function for the largest possible number of applications, for reasons of cost-effectiveness and (2) the antenna is connected to a chip which, when in operation, varies its input impedance between two values.

Based on this, a question arises for which antenna impedance value must be taken into account in order to optimize communication between the reader and the tag. We will see that the answer is not so obvious, and it becomes even less clear when we consider that the two impedance states of the chip are dependent on the frequency and the power received. The answer to this question depends largely on the intended application. Likewise, and this is the most interesting aspect, work on this point is still in progress. Indeed, numerous studies worldwide are focused on the development of this technology, which remains highly promising, that is with extremely high development potential.

If we return to the "antenna" aspect, we can see that the design of antennas compatible with the expected applications in the domain of traceability arouses a great deal of interest in the academic and industrial sectors. Miniature, low-cost, relatively broadband tag antennas currently represent a significant challenge in the field of UHF RFID systems. In practice, UHF RFID tags must function no matter what the usage environment is. As we have already noted, for economic reasons, a single tag must be able to be used no matter what item is needed to be tracked, and no matter where it is located. These parameters are highly variable, and are theoretically unknown during the tag design phase; moreover, they have an extremely important impact on performance, and must therefore be taken into account in one way or another during the design of RFID tag antennas. Indeed, the high pressure currently weighing on RFID has had a direct impact on the antenna design methods used up to now. As we will see, design approaches must make it possible to obtain tags that will be resistant to various environments. Finally, practical and economic constraints make both small dimensions and a wide operating frequency band necessary. Research in this area is extensive, aimed most often at developing a "universal" tag, which remains more illusion than reality at this point. To do this, as we will see in detail later on, the development of automated design approaches taking as many stresses as possible into consideration seems to be a vital element in the eventual success of this research. In the meantime, let us return to the different RFID technologies.

As noted above, RFID is a major technology in the field of identification. Its field of action is in a perpetual state of evolution. The first article describing the operating principle of RFID was written by Stockman in 1947 [STO 48]. The first practical application was the Identification Friend or Foe (IFF) system used during World War II to recognize Allied planes. Commercially speaking, the Electronic Article Surveillance (EAS) system was the first one developed in the 1960s and subsequently put on the market [FIN 10]. However, a fairly significant period of time was required after the appearance of the first RFID article for the technology to reach the applicative sector, that is, for it to reach a level of development sufficient for large-scale deployment.

The main advantage of RFID is its ability to automate ID capture procedures. This reduces human interaction so as to limit input errors, and increases the speed of information capture. RFID is used in thousands of applications; today, these are mainly used in the sectors of security (access control), traceability and identification, and it will soon be used in low-cost wireless sensors and short-range localization systems. From this, it is clear that each application includes its own constraints, which RFID must take into consideration. These constraints can be extremely diverse, concerning tag dimensions, costs, reliability, security, etc.; and they may even be contradictory from one application to the next. Taking the example of reading distance, it is clear that for security applications (access badges), a short distance is preferable, while for large retailers longer reading distances are needed to conduct inventories. Thus, all of these applications generate a considerable number of constraints, which have resulted in the application of not one RFID technology, but a large number of them. This is why, if we look more closely at RFID tags (Figure 1.2), we can see how widely they differ from each other. Consequently, we speak of families of RFID tags, which are necessarily composed of subfamilies, and so forth.

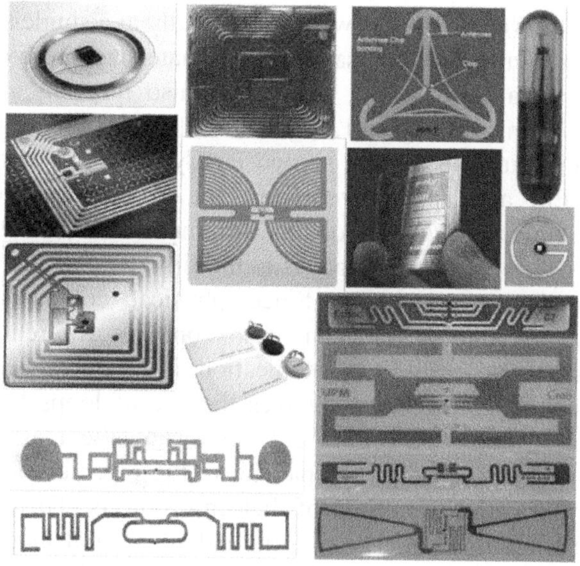

Figure 1.2. *Different types of RFID tags for various applications. Tags are not all shown on the same scale*

All the specifications introduced by applications have led designers to use several frequency bands. The use of the electromagnetic (EM) spectrum is widely restricted by governments, in terms of both frequency and power; therefore, RFID is confined to different frequency bands, sometimes very narrow ones, which do not require a license to be used. Figure 1.3 shows these different frequency bands (central frequency and associated wavelength). We can also see in this figure that an important distinction in terms of communication must be made. Two major families of RFID tags are clearly evident; the simplest way to introduce these is to go back to the applicative level. Here, we are seeking to identify the objects surrounding us, that is, mainly objects with dimensions ranging from a few centimeters to several meters. For practical reasons, tags must almost always be as small as possible, with dimensions comparable to the smallest objects being tagged. Based on this, certain standards are observed with sizes more or less comparable to that of a credit card. If we look at the wavelengths associated with the frequencies used in RFID (Figure 1.3), we can see a very wide range, from 5 cm to 2.4 km.

Consequently, we note that if we look generally at a single size for all antennas (or in any case, the same approximate size, more or less the dimensions of a credit card), these must also function in different operating systems. As Figure 1.4 shows, we have a first family of tags operating in near fields, and a second family of tags operating in far fields. In EM and applicative terms, the difference is extremely important. It enables us to obtain different reading distances: in contact or over very short distances in the first case, and over several meters in the second case. Since physical communication is completely different, we can also obtain different reading characteristics. Near-field communication is often carried out via magnetic coupling, and thus involves the use of loops by both the reader and the tag (see Figure 1.4). By using the magnetic field, reading is less disrupted in comparison to the external environment, which is most often composed of dielectric objects. To improve the extent of the reading zone, a uniform and substantial magnetic field must be transmitted throughout the zone. For far-field communication, the tag is no longer composed of a loop, but rather of an antenna (see Figure 1.4) able to receive and transmit an EM wave. Note that, to ensure far-field communications, if we place ourselves at around 900 MHz, using antennas with classic dimensions of $\lambda/2$, a minimum distance between the reader and the tag on the order of 20–30 cm is required. In practice, identification in this close-field zone (0–20 to 30 cm) must still be able to do this. For example, at the end of a tag assembly line, a manufacturer must encode its tags with a unique ID and verify that they are working properly. For practical reasons, this operation must take place with virtual contact occurring between the tag and the encoder. This is part of the reason why the vast majority of passive UHF tags include a loop around the chip, to facilitate this operation, that is, to guarantee close-field communication (see Figure 1.2).

Through these few examples, we can see how there is not a single RFID technology, rather multiple technologies that can be very different from each other. This makes RFID, from all dimensions, a subject for which an exhaustive discussion is a highly ambitious undertaking. However, this is not the objective of this book. For further details, we refer the readers to various books [BOL 10,

DOB 07, FIN 10, LAH 14, LEH 12, PAR 09]. It is also important to delineate our field of study clearly. Figure 1.5 shows one classification of families of RFID tags.

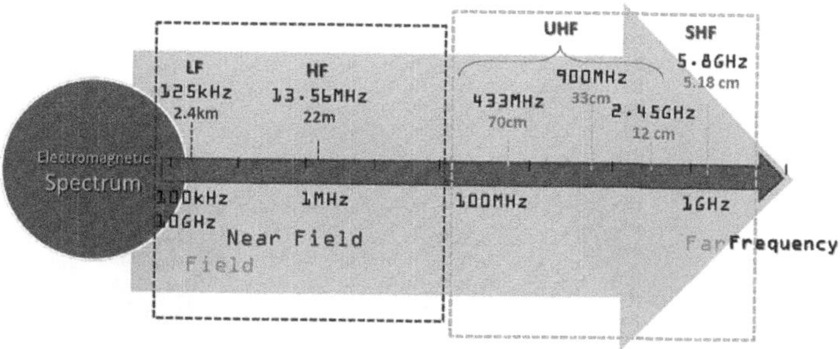

Figure 1.3. *Principal frequency bands used by RFID*

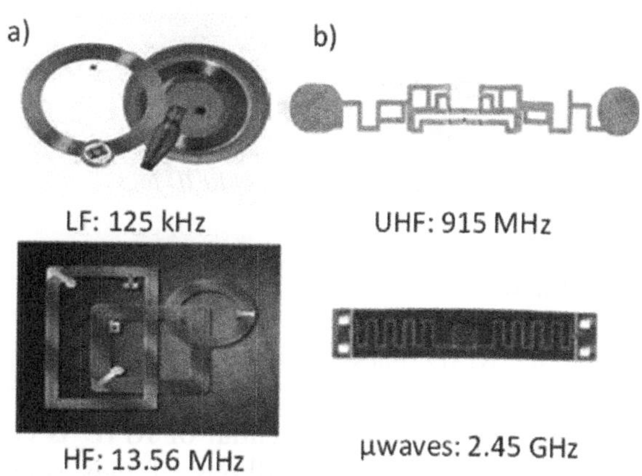

Figure 1.4. *Different types of RFID tags. a) Tags with loop "antenna" functioning in near field via magnetic coupling. b) Tags with "antenna", functioning in far field, that is, able to receive and transmit an EM wave. Tags are not all shown on the same scale*

We can define six main categories, characterized by chip feed, read range, information support, functionalities offered and programming used. We will, henceforth, focus mainly on the hardware considerations, in other words, on the first four fields. Thus, our subject of study will involve mainly passive tags (which have no battery, and in which the power supply comes from the reader only), which are able to communicate from several meters away from the reader, with or without a chip. We will also pinpoint the keys used to divert the primary function of an RFID tag – identification – in order to add functions of authentication, localization or sensor.

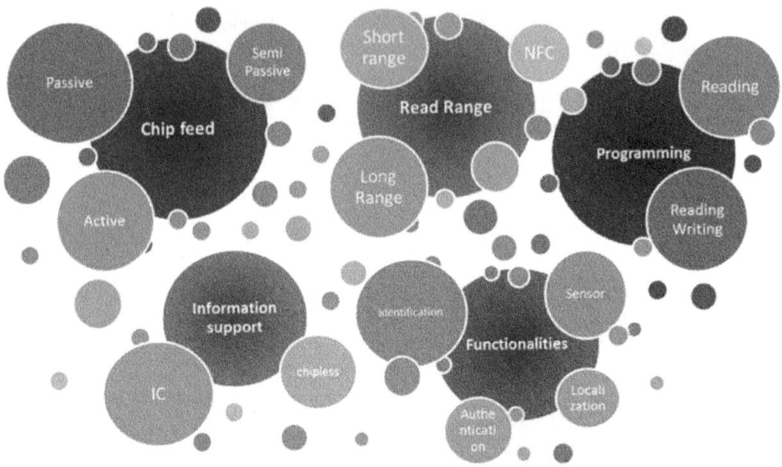

Figure 1.5. *Classification of RFID tags*

1.2. The RFID market

If we examine the RFID market, numerous studies have shown that RFID identification will experience a significant economic boom in the years to come. The economic potential of RFID is considerable, with billions of euros and tens of billions of tags sold each year [IDT 14]. Additionally, as shown by numerous prospective studies, the field of passive RFID has experienced double-figure annual advancement for several years, and this progression will continue during the current decade. Thus, it is already a considerable and fast-growing market, and it is expected that the use of RFID tags in many areas of daily life will become both widespread and common in the future. One concrete example of this generalization is the tagging of cardboard and paper packaging materials. It is estimated that 320 billion cardboard/paper packaging units are sold annually in Europe, a figure that represents only 20% of the overall total packaging materials of every type (cardboard, paper, glass and plastic). It is clear that the number of tags may rise into the hundreds of billions annually. Likewise, for many applications we are moving toward the use of tags as simple disposable consumables.

Numerous prospective studies are available on this subject [IDT 14], which generally predict the 10-year evolution of the RFID market. These predictions clearly indicate that quasi-exponential growth is expected, from a market estimated at 9 billion dollars in 2014 (or 7 billion tags sold) to more than 30 billion dollars in 2024. We would point out that even though these figures are very large in terms of numbers of tags, they are nothing in comparison to the number of barcodes used each year, which is estimated to be between 5 and 6 thousand billion. In 2012, the number of UHF tags sold per year (mostly passive tags) exceeded the number of high frequency (HF) tags. However, value-wise the UHF RFID market remains weak (on the order of 10%) compared to the HF RFID market.

Some studies also make it possible to separate the distribution between chipless tags and classic-chipped tags. The market proportion of chipless tags is currently negligible (0.1% in 2012), but this trend is expected to progress steeply [IDT 14]; the 2019 proportion of chipless tags on the market is expected to reach 31%. If we consider the number of tags sold per year, the figures are even more evocative; in fact, chipless tags – which represented 0.5% of tags in 2012 – should reach 80% in 2019. It is believed that the chipless boom will be driven principally by the development of printed electronics. In managing traceability issues, the cost of the solution, and particularly that of tags, is of primary importance, and this pressure is even greater since the objects to be tracked have little unitary value but exist in large numbers. Expectations for these applications, situated at the opposite end of the spectrum from the luxury products sector, continue to rise. It is clear, then, that cost constraints are a fundamental given in RFID. This is why the development of low-cost tags is eagerly awaited. Likewise, the possibility of being able to manufacture new families of tags using standard paper-industry techniques is a definite advantage, and the very reason for the recent development of chipless RFID. The cost of a UHF tag can be divided roughly equally between the price of the materials (conductive parts composing the antenna; dielectric parts for the medium); the price of the chip; and the price of the chip's report on the antenna. Therefore, eliminating the chip reduces the price of a tag by at least two-thirds. We will see in Part Two of this book, dedicated to chipless RFID, that this technology is compatible with mass-production processes currently used in the paper-industry. It has been shown that it is possible to produce tags of this type for a unit cost on the order of 0.4 euro cents, that is, for the same approximate price as that of barcodes, or around 20 times less than a passive UHF RFID tag. We might suggest that the desired objective is 1 American cent in 2018 for several hundred billion tags sold across all categories (chipped, chipless, passive, active, etc.). If we examine this figure in detail, we can see that the average price of chipped tags is estimated at 4 American cents, while that of chipless tags is 0.4 cents. Always with the goal of reducing the cost of tags, one solution might consist of printing the tag directly on the product or packaging material to be identified. In this case, the tag cost could quickly drop by a factor of 10.

1.3. Issues in RFID

As we have seen, RFID is an extremely vast domain with a large number of applications, each of which generates its own constraints. In this section, we will look back over the principal limitations of passive UHF RFID and examine the research that has been conducted to rectify them. This research will be discussed in more detail in later chapters.

1.3.1. *Robustness of reading*

1.3.1.1. *Description of problem*

The use of passive UHF RFID tags is becoming increasingly common. The intensive development of UHF RFID has been slowed down, however, for several reasons. The main technical reason for this is the unreliability of tags, particularly when they are in an environment other than the one for which they were specifically designed [UKK 06]. In reality, the context within which UHF RFID tags are used has a significant effect on their performances. No universal-use tag currently exists, that is a tag that is guaranteed to function in any environment. In terms of the use of a tag, the directly disruptive element is nothing but the object on which this tag is placed [RAO 05]. For this technology to be economically viable, it is important that a tag is able to be used with the largest possible number of objects.

The problem of robustness of reading is undoubtedly the most important technical difficulty encountered by UHF RFID, and the field of application of this technology is truly limited as a result of it. By "robustness of reading", we mean the system's capacity to recover a tag's ID independently of the environment in which the tag and the reader are located. We can characterize this concept more precisely through the reading rate of tags in a given environment. This point is critical, as it is directly related to the use of RF waves for communication; this is in addition to the extremely strict limitations imposed by industry. The benchmark reference in the matter is none other than the barcode, which yields a very high reading rate, one that RFID must at least equal. In the world of logistics, 99% reading rates are most often inadequate. One reading failure in 1,000 is a more desirable value. However, these reading rates are difficult to achieve, and impossible for passive UHF RFID, which explains why UHF RFID is rare if ever used in the large retail sector. This is why the use of other technologies, such as HF RFID, which operates by the inductive coupling principle, is more suitable in this case.

If we consider passive HF RFID tags, we can see that they are relatively similar to UHF tags in terms of manufacturing and materials used to make them. They are also composed of a chip and metal strips ensuring communication with the reader. In the end, the costs of both are similar, although HF RFID tags are generally more costly because they require the use of an intermediary to link the two ends of the loop to the chip. The area of use is different, in that HF tags work for short-distance applications, with a maximum reading distance of a few dozen centimeters. Despite this limitation in distance, HF tags are currently most frequently used in a considerable number of sectors, particularly mass-market retailing. The reason for this is not an economic one, but rather a technical one. In real situations of use, HF tags are more robust than UHF tags. This can be explained theoretically: in near-field zones we note that, unlike the electric field, which is highly sensitive to dielectric materials and metals, the magnetic field is much less sensitive to these materials, which mostly make up objects used in everyday life. In this case, communication is accomplished via inductive coupling, in which the antennas best suited for this type of transmission are simply loops. In applicative terms, this mode of communication has an extremely significant impact for the user. A single HF tag can be used without previous testing to tag a large number of objects of various types. For example, HF RFID system operations will hardly be disrupted whether the object being tracked contains metallic inks, which are used in some packaging materials, or not. In UHF, the tag's external environment (both the object in contact that is being tagged and the near surrounding environment) may negatively affect system performance rapidly, thus limiting the field of application of this technology. These simple observations show clearly to what extent a reading rate on the order of 99% represents a real challenge. It should be noted, however, that HF RFID, though more favorable in this sense, has experienced numerous failures in the large retail distribution field.

1.3.1.2. *Solution contributed*

We will start, then, from the observation that objects being tagged have EM properties and geometric dimensions that are potentially different. We may add to this point that, in applicative terms, it is increasingly desirable to be able to integrate the tag directly into an object to be traced. It is of particular interest to be able to track the object throughout the phases of its production, and then all along the distribution chain until it reaches the final customer, for example. For all these reasons, specific work focused on the development of a new design methodology for RFID tags has been conducted with the aim of promoting the mass deployment of passive UHF RFID technology [CHA 09, CHA 10a, CHA 11a, CHA 11b]. Chapter 2 describes the development of software capable of automatically designing the tag based directly on technical specifications [PER 10]. Indications entered by the user may be limited to the geometric dimensions of the final tag and the electrical properties of the RFID chip used. If the user so desires, he or she can also provide information about the environment in which the tag will be used (range of variation of the permittivity of various tag supports; presence of metal or lack thereof, etc.). The application then sends back the tag antenna form that is compliant with the limitations imposed. This approach is radically different from the classic antenna design solutions for UHF RFID tags, which are based on a purely empirical design approach based above all on the principles of folding wire antennas and current loops (particularly for near-field aspects). This new design approach makes it possible to take usage requirements into consideration, and these constraints play a role in determining the shape of the antenna. Thus an antenna's topology is not imposed theoretically, as is the case in classically used methods. In this, the approach developed does not simply consist of optimizing a topology that has already been chosen, but rather of designing the antenna from start to finish on the basis of the constraints required for it. To do this, EM simulation software (CST, Ansoft Designer, etc.) and calculation software (e.g. Matlab) have been used in combination [CHA 10a, CHA 10b, CHA 11a]. In order to comply with the constraints imposed during the design of an antenna for a UHF RFID tag, an optimization process is used based on the concept of genetic algorithms (GAs). The principle of optimization is based on an iterative process that consists of generating an antenna form, simulating it and evaluating it in relation to the constraints imposed. The generation of forms from one iteration to another is done on the basis of an evolutionist principle. This is repeated until an antenna design is obtained that best satisfies the specifications. This approach is highly flexible, and makes it possible to take various constraints into account throughout the entire design phase, particularly in terms of the tag's external environment.

1.3.2. *Tag prices*

The question of cost is omnipresent in RFID. Like technologies meant to be used by the general public, this dimension is particularly present in RFID, where tags are often intended to be disposable. Here again, the barcode is the benchmark reference in the matter. The lower the cost of a tag, the more it can be used on low-value objects, and thus in greater numbers. This cost constraint is also present for readers, where the price cannot exceed a few hundred euros per unit.

In RFID, cost determines the production techniques and materials used in manufacturing tags. As an example, the most commonly used substrate is polyethylene terephthalate (PET), which is a very common plastic material. In order to reduce the cost of tags and thus to move toward the projections discussed above, a simple solution consists of producing chipless tags. In order to better understand the motivations behind the development of this approach, it is of interest to compare it to the barcode. Note that RFID is still a relatively complex identification technology compared to barcodes; the latter are extremely simple to apply and use. They are also completely standardized and universal in their operating principle. In addition, barcodes are extremely low-cost, both for tags and readers. For example, anyone can, using free software, generate the barcode corresponding to his or her own ID and print it using a standard office printer. This code can be read using a webcam or a smartphone. However, the main disadvantage of this technology is the way in which it captures information, which requires human intervention. If we evaluate RFID in the same way, we can see that the advantages of one of these technologies will appear as the disadvantages of the other. In fact, communication through radio waves is the principal benefit of RFID, enabling as it does the automated capture of an ID. This flexibility of reading is accompanied by the ability to execute multiple readings, and it is also possible to read over substantial distances. However, the RFID solution is not simple, and it requires the use of an electronic chip and a communication protocol that adds significant cost for tags and readers and can cause mechanical and reading robustness difficulties. Also, this solution is not universal, insofar as it uses frequency bands that may differ from one region to another, and this has an effect on the intrinsic performances of RFID systems in the end. All of these observations explain why more and more research is aimed at developing new traceability systems. Of these, chipless RFID, still referred to as an "RF barcode", seems highly promising [PER 11, PRE 10, TED 09, TED 10a, TED 10b]. As we will see, it falls in applicative terms somewhere between classic

RFID and barcodes, that is, where the guiding objective needs to combine the advantages of RFID with those of barcodes.

With regard to the issue of reader-side cost, the current trend is toward readers with "agile" antennas, that is, with significant gains as well as the ability to turn off the beam [LEE 13]. In fact, by increasing antenna gain, it is possible to reduce significantly the price of reader amplification stages.

1.3.3. *From identification toward sensor function*

One of RFID's strengths lies in the fact that this technology is widespread and is regulated by clearly established norms. Based on this, the temptation is great to try to reuse this technology by simply adding functionalities to it without modifying it in depth. RFID can be seen as a communication system, and the objective is to add other functions to its basic identification function. Indeed, rather than seeking to make tags more environment-resistant within the context of classic RFID applications, it is possible to try to make it sensitive to a physical value so as to be able to measure it. Any phenomenon that will modify the adaptation of the chip to the antenna, and thus the modulation of the backscattered signal, can therefore be used.

As an example, it has been demonstrated that it is possible to measure the level of a liquid in a container, or to localize it. Here, we can see the full interest of this approach, in that no major modification of the RFID system is required. In simple terms, the idea is to take a reader and commercial tags and to process the information recovered by the reader in such a way as to go back to the physical value being measured. In this way, we obtain a sensor function associated with a unique ID, all relying on a communication system that is already widespread. To optimize the sensitivity of the tag, it is possible to design a specific tag antenna for the application geared to a given chip. These ideas will be discussed in Chapter 3. It is true that the "reuse" of UHF RFID as a sensor represents a major challenge for this technology, and has excited a great deal of enthusiasm in the scientific community.

1.4. Conclusion

In this introductory chapter, we have described the principal issues that this book will address. The observations thus evoked have made it possible to introduce chipless RFID, which will play an important role later in the book. First, though, we will examine in detail the question of the design of a passive UHF RFID tag antenna. The next chapter is devoted entirely to this subject. We will see how it is possible to approach this issue from a new angle, and thus lead into an automated antenna design methodology for RFID tags.

1.5. Bibliography

[BOL 10] BOLIC M., SIMPLOT-RYL D., STOJMENOVIC I., *RFID Systems: Research Trends and Challenges*, Wiley, 2010.

[CHA 09] CHAABANE H., PERRET E., TEDJINI S., "Conception automatisée d'un tag RFID UHF robuste", *16èmes Journées Nationales Micro-Ondes*, Grenoble (France), 2009.

[CHA 10a] CHAABANE H., PERRET E., TEDJINI S., "Towards design of robust UHF RFID tag", *4th IEEE RFID Conference*, Florida, 14–16 April 2010.

[CHA 10b] CHAABANE H., PERRET E., TEDJINI S., "Computer aided-design for robust UHF RFID antenna", *Wireless Systems International Meeting (WSIM '10)*, Campina Grande, Brazil, 2010.

[CHA 11a] CHAABANE H., PERRET E., TEDJINI S., "A methodology for the design of frequency and environment robust UHF RFID tags", *IEEE Transactions on Antennas and Propagation*, vol. 59, no. 9, pp. 3436–3441, 2011.

[CHA 11b] CHAABANE H., PERRET E., DAIKI M., *et al.*, "RFID tag design approach based on chip multi states impedance", *Asia-Pacific Microwave Conference Proceedings (APMC)*, Melbourne, pp. 1466–1469, 5–8 December 2011.

[DOB 07] DOBKIN D.M., *The RF in RFID: Passive UHF RFID in Practice*, Elsevier Science, 2007.

[FIN 10] FINKENZELLER K., *RFID Handbook: Fundamentals and Applications in Contactless Smart Cards, Radio Frequency Identification and Near-Field Communication,* Wiley, 2010.

[IDT 14] IDTECHEX, 2014. Available: http://www.idtechex.com/.

[LAH 14] LAHEURTE J.M., RIPOLL C., PARET D., *et al.*, *UHF RFID Technologies for Identification and Traceability*, Wiley, 2014.

[LEE 13] LEE W.-S., KHANG S.-T., LEE W.-S., *et al.*, "Wide-coverage array antenna using a dual-beam switching for UHF RFID applications", *2013 IEEE International Conference on RFID*, Orlando, FL, 2013.

[LEH 12] LEHPAMER H., *RFID Design Principles*, Artech House, 2012.

[PAR 09] PARET D., *RFID at Ultra and Super High Frequencies: Theory and Application*, Wiley, 2009.

[PER 10] PERRET E., CHAABANE H., TEDJINI S., "RFIDTool", *Dépôt du logiciel auprès de L'Agence pour la Protection des Programmes (APP) Inter Deposit Digital Number: IDDN.FR.001.390041.000S.C. 2010.000.20600,* October 2010.

[PER 11] PERRET E., HAMDI M., VENA A., *et al.*, "RF and THz identification using a new generation of chipless RFID tags", *Radioengineering – Special Issue: Emerging Materials, Methods, and Technologies in Antenna & Propagation,* vol. 20, no. 2, pp. 380–386, June 2011.

[PRE 10] PRERADOVIC S., KARMAKAR N.C., "Chipless RFID: bar code of the future", *IEEE Microwave Magazine*, vol. 11, no. 7, pp. 87–97, 2010.

[RAO 05] RAO K.V.S., NIKITIN P.V., LAM S.F., "Antenna design for UHF RFID tags: a review and a practical application", *IEEE Transactions on Antennas and Propagation*, vol. 53, no. 42, pp. 3870–3876, 2005.

[STO 48] STOCKMAN H., "Communication by means of reflected power", *Proceedings of the IRE*, vol. 36, no. 10, pp. 1196–1204, 1948.

[TED 09] TEDJINI S., PERRET E., "Radio-frequency identification systems and advances in tag design", URSI *Radio Science Bulletin,* vol. 331, pp. 9–20, December 2009.

[TED 10a] TEDJINI S., PERRET E., DEEPU V., *et al.*, "The internet of things: 20th tyrrhenian workshop on digital communication", *Chipless Tags, the Next RFID Frontier*, DANIEL G., ANTONIO I., GIACOMO M., *et al.* (eds.), Springer New York, pp. 239–249, 2010.

[TED 10b] TEDJINI S., PERRET E., DEEPU V., *et al.*, "Chipless tags for RF and THz identification", *4th European Conference on Antennas and Propagation (EuCAP '10)*, Barcelona, Spain, pp. 1–5, 2010.

[UKK 06] UKKONEN L., SCHAFFRATH M., KATAJA J., *et al.*, "Evolutionary RFID tag antenna design for paper industry applications", *International Journal of Radio Frequency Identification Technology and Applications*, vol. 1, no. 1, pp. 107–122, 2006.

2

Antenna Design for UHF RFID Tags

2.1. Introduction

Radio frequency identification (RFID) is a major technology which, for more than a decade now, has undergone significant development in terms of applications. It is a technique for capturing information from a tag containing data, via remote radio reading. The label is composed of an electronic chip and an antenna that enables communication with a dedicated reader (Figure 2.1). As we saw in Chapter 1, despite the many advantages of RFID, its deployment has been hindered by multiple factors: economic as well as technological and societal. Of these difficulties, we have already noted the cost of tags, which is still too high, and the lack of reliability. We might also add the lack of security of the information contained in the RFID chip, and factors involving the protection of private life, health precautions and the recycling of tags (which number in the hundreds of billions per year).

2.2. Essential RFID parameters

The first task is to distinguish high frequency (HF) tags from ultra high frequency (UHF) tags in terms of design approach. For HF tags and readers, the metal strips connected to the chip do not constitute an antenna, properly speaking, but rather a coil. The principle of information transmission relies in this case on the variation of the reactive energy around the coil. The objective, then, is to maximize the coupling between the "antennas" of the tag and the reader. The task of design consists of adjusting the size of the loop (total length, widths of strips, number of turns, spacing between strips, etc.) in order to cause the circuit (coil plus chip) to resonate at the desired frequency. The coil is modeled by an R, L, C equivalent electric circuit, and the electrical characteristics of the chip are supplied by the manufacturer. Switching between geometrical and electrical values is done using analytical formulas that are most often empirical. This approach makes it possible quickly to establish the initial dimensions of the circuit. An optimization phase with an electromagnetic (EM) simulator may complete the design phase. Compared to UHF tags, the fact of working at lower frequencies and in close field allows us to obtain robust tag/reader systems and facilitates the full-scale deployment of this approach. HF RFID has been a mature technology for several years now, and its applicative advantages and limits are well known.

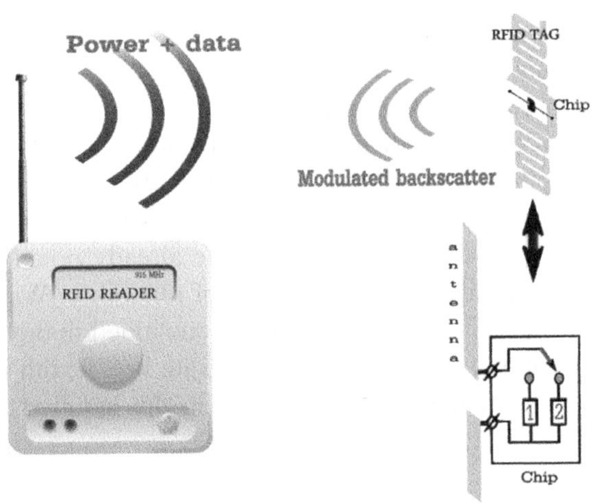

Figure 2.1. *Schematic diagram of a passive UHF RFID system*

The design of UHF tags is quite different for several reasons. For example, in UHF there are no analytical formulas for moving from geometric parameters to the equivalent electrical model. This fact is representative of the degree of complexity of the UHF design phase.

Next, we will focus on issues that are proper to the design of UHF tags, and more particular still on passive tags. The antenna is the component vital to the chip's ability to communicate with the reader. Antenna design is at the true heart of the issues with RFID. The antenna must make it possible to collect enough energy to power the chip, while also radiating energy back toward the reader [NIK 06, RAO 05a]. Note also that optimizing the transfer of power from the antenna toward the chip necessarily results in a reduction of the reverse transfer of power, that is, the transmission of energy from the antenna to the reader. This is the base of the coding principle used for information exchange.

For communication to take place between a tag and a reader, two conditions must be satisfied. The first condition consists of ensuring that the power arriving in the chip, which we will write as P_{chip}, is high enough, that is, higher than the minimum power required to activate the chip. This latter parameter a "builder" value inherent in the chip, and we will denote it as P_{th}, for *threshold power*. However, the fact of the chip being powered (and thus sending its identification signal back in the direction of the reader) is not enough to ensure the correct operation of the tag/reader system. The second condition that must be met for the message to be intelligible to the reader is that the two coding states (the values 0 and 1 in baseband) received by the reader must be sufficiently differentiated. For this to be the case, in practice, the ΔRCS[1] measurement – that is, the difference of the radar cross-section (RCS), between each of the states – provides information about the robustness of the communication [FIN 10, PAR 09, PER 12]. The differentiation between these two states, as we will see, depends on a number of parameters that are more significant only when we try to satisfy the first condition (power supply to the chip). The main difference lies in the fact that in order to characterize this modulation it is necessary to know the two impedance states of the chip, only the "adapted" state is required to maximize the power received by the chip.

---

1 $\Delta RCS = RCS^0 - RCS^1$, where RCS^0 and RCS^1 correspond, respectively, to the RCS values of the tag loaded by Z_c^0, Z_c^1.

2.2.1. *Putting into equation of reader-tag links*

A tag combines an antenna and a chip, which is an integrated circuit (IC). As shown in Figure 2.2, each of these two components can be modeled by a complex impedance, which we will write as Z_a and Z_c for the antenna and the chip, respectively. Communication relies on the principle of load modulation. When the RFID tag receives the modulated signal from the reader with sufficient power (descending link: from reader to tag), the chip varies its impedance between two states, written here as Z_c^0, Z_c^1 (with $Z_c^i = R_c^i + jX_c^i$. i=[0, 1]). Thus, for each impedance state, the RFID tag presents a specific RCS. The signal generated by the tag is the reflection of the wave (sent continuously by the reader) off the tag's antenna, loaded with an impedance that presents two different values. This is how the return signal is obtained, that is, via the rising link from the tag to the reader. Figure 2.3 shows the capture via oscilloscope of an RFID frame. The first part of the signal is the request sent by the reader (here an arbitrary waveform generator is used to emulate the RFID reader), characterized by a distinct front. On the order of 50 μs after the end of the request from the reader, we can see the signal backscattered by the tag. It is characterized by a variation that is low in amplitude compared to the one generated by the reader (on the order here of 0.1 V compared to 1 V). We can see that the first part of this signal is composed of regularly spaced gaps (successions of 0 and 1), while the second part is simply the tag identifier (irregular gaps).

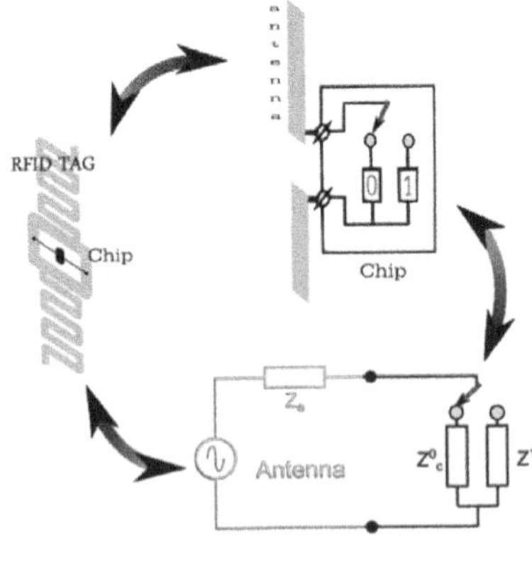

Figure 2.2. *Equivalent electrical schema of a passive UHF RFID tag*

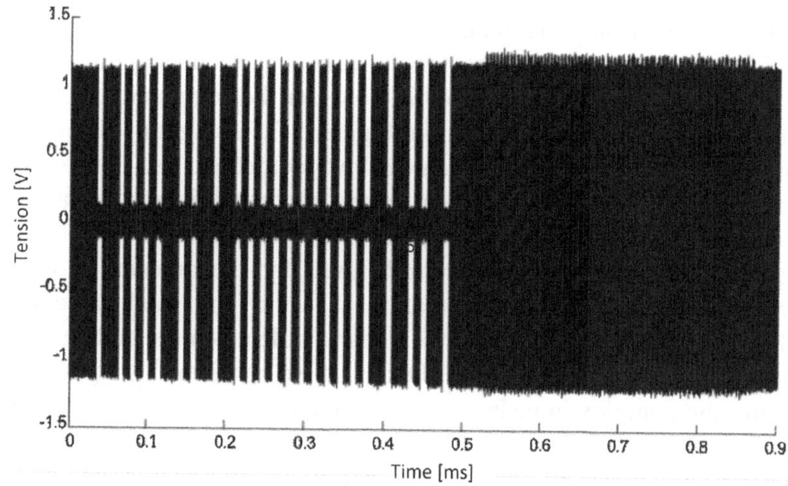

Figure 2.3. *Oscilloscope measurement of an RFID frame. We can see the reader request (left-hand side) and the tag's response (right-hand side, which is of very low amplitude compared to the reader request)*

Let us look more closely at the two distinct impedances of the chip. If we want to maximize the power available to the IC in order to ensure an adequate power supply, we must – for one of the two impedances – carry out a complex conjugate matching with the impedance of the antenna [FIN 10, PAR 09, RAO 05b]. This characterizes the "absorbent" condition required for the downlink. If we look now at the uplink, in order for the signal to be intelligible to the reader, that is, for the two symbols to be as different as possible, the two impedances must yield two different backscattered waves reflected by the tag (containing two states on the amplitude and/or phase) that are as differentiative as possible. Thus, we can see that the reader receives two distinct signals that will be used to code information. The matching condition of power to optimize the downlink is, therefore, $Z_a = Z_c^{0\,*}$, where Z_c^0 is the chip impedance corresponding to the "absorbent" state. This is why this condition is very important, even central in passive UHF RFID. In practice, to obtain this condition, it is not possible, for reasons of cost, to use an external matching circuit. Only an adjustment of the antenna, i.e. the molding of its geometry, can be used to obtain it.

Before describing the main principles of RFID tag design, we will begin by introducing the characteristic values of the up and downlinks of RFID communication. To do this, consider the equivalent electrical schema of the antenna shown in Figure 2.2. Using this Thevenin equivalent circuit, we can calculate a link budget and use it to define the principal parameters in which we are interested. It is important to note that by using this equivalent electrical schema, we are implicitly assuming that the EM field retrodiffused by the antenna in open circuit is zero. This approximation is valid for small antennas [GRE 63, HAR 63], which are like linear dipole antennas, and thus the antennas used most often for RFID tags. If we consider this equivalent circuit, the complex impedances of the antenna and the chip are given by $Z_a = R_a + jX_a$ and $Z_c^i = R_c^i + jX_c^i$, respectively (with i=[0, 1]). In addition, the complex chip impedance is generally dependent on frequency and input power (i.e. Zic(f,p)). However, it is possible to express, depending on the impedances Z_c^i and Z_a, the ratio between the total power reaching the tag P_{tag} and the power actually reaching the chip P_{chip} This ratio is the power transmission coefficient τ^i. All of the calculations shown below are summarized in Figure 2.4; the following relationships will be considered [NIK 05, RAO 05a]:

$$P_{chip} = \tau^i \, P_{tag},$$

with:

$$\tau^i = 1 - \left|\Gamma_{tag}^i\right|^2,$$

$$\Gamma_{tag}^i = \frac{Z_c^i - Z_a^*}{Z_c^i + Z_a} \qquad [2.1]$$

or, depending on the impedances:

$$\tau^i = \frac{4R_a R_c^i}{\left|Z_c^i + Z_a\right|^2}$$

The increase of τ^i (a value between 1 and 0; note also that it brings in only a single Z_c^i at a time[2]) is accompanied by an increase of the

2 The matching condition is fulfilled at load $i = 0$, which we will write as: $\tau^i = \tau^0 = \tau$.

power reaching the chip and makes it possible to increase the distance d_{up}, which corresponds to the maximum distance at which the tag receives enough energy to be activated and thus for its load to be modulated. In this zone, the power P_{chip} is greater than P_{th}, which is the minimum power necessary to supply the RFID chip.

Now, let us look at the downlink; we are seeking here to maximize the modulation efficiency so as to improve the robustness of communication. Similarly, the link between the retrodiffused power available to the reader P_{rec} and the power in the tag can be written as: $P_{rec} = K^i P_{tag}$ where K^i is the backscattered coefficient. It can also be expressed in terms of the values of the two impedances: $K^i = \frac{4R_a^2}{|Z_c^i+Z_a|^2}$.

This result is obtained assuming once again that the EM field retrodiffused by the tag antenna in open circuit is negligible [SKA 09]. Based on this, we can express the difference in power which can be caught by the reader and pertaining to each of the two impedance states: $\Delta P_{rec} = (K^0 - K^1)P_{tag}$.

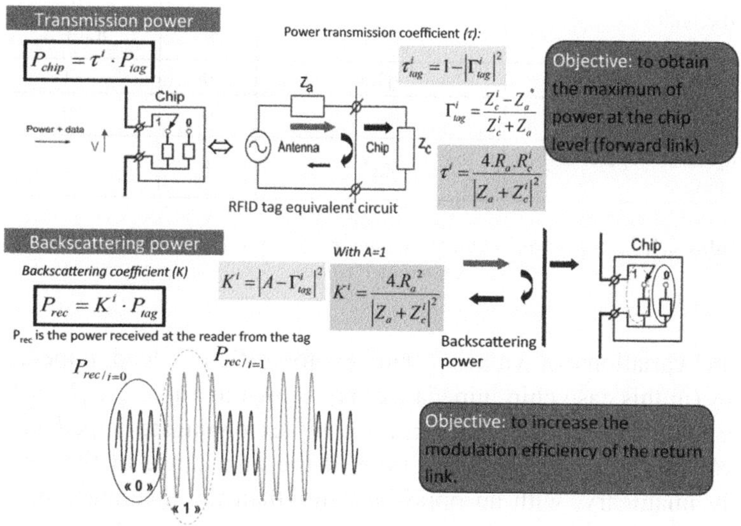

Figure 2.4. *Schematic and analytical presentation of the values involved in the description of the up- and downlinks of an RFID system*

If we consider that the information is coded in amplitude (amplitude modulation) and that the reader sensitivity is Δρ, we can write the condition for which the reader is able to collect information from the tag as $\Delta P_{rec} > \Delta\rho$.

Since ΔP_{rec} is a function of the distance between the tag and the reader, the previous expression is used to determine the maximum distance d_{down} over which the reader is able to decode the information from the tag. Note that, unlike the uplink, this distance depends on the two impedance states of the chip. Moreover, we can see clearly that depending on the link taken into account (up or downlink), we obtain two characteristic two distances, which are different from each other but equally independent. The maximum reading distance *r* (a value usually called the *read range*) is generally the minimal value between d_{down} and d_{up}.

$$r = \min(d_{down}, d_{up}) \qquad [2.2]$$

	Zc	τ	K^i
Short circuit	0	0	$\frac{4R_a^2}{R_a^2+X_a^2}$
Re(Zc)=0	-jXa	0	4
Open circuit	∞	0	0
Complex conjugate matching	Za*	1	1

Table 2.1. *Variations of coefficients τ and K^i for different characteristic load impedance values*

The variations of values τ and K^i for different load impedance values (in this case chip impedance) are shown in Table 2.1 [NIK 06]. When the antenna is loaded with a complex conjugate impedance, τ and K^i are equal. At the same time, if the impedance of the load is purely imaginary, with an opposite value than the imaginary part of the antenna impedance, the short-circuited antenna reflects an amount of power equal to four times that of the conjugated case. This is the case in which the backscattered power is maximal. It is clear that the transmission coefficient τ and the difference of the reflection coefficients $K^0 - K^1$ cannot be maximal at the same time for a given load impedance.

Looking at this more closely, and focusing more precisely on the significant parameter ΔP_{rec}, it has been shown that a compromise must be found between these two coefficients, through the choice of antenna impedance [BOL 10, CHA 11a]. In this case, we will assume that the chip has been chosen and that its impedance is known. This also means concretely that the condition $Z_a = Z_c^{0\,*}$ is not necessarily the condition that maximizes the read range [2.2] for a given chip. However, in passive UHF RFID, reader sensitivity is good enough for the limiting condition to be that of the uplink (in this case, we have $r = \min(d_{down}, d_{up}) = d_{up}$). Indeed, unlike the tags, the readers are supplied with power. This means we can assume that if the tag receives enough power to be activated, the reader will be able to collect the information delivered by the tag. In this case, the read range is solely a function of Z_c^0, and so we are seeking to maximize τ, and thus to create a condition of complex conjugate matching on the whole frequency band being considered. In summary, with this hypothesis we assume that simply activating the tag will result in communication being established. Subsequently, the radar equation enables us to write the read range in terms of the other parameters of both the tag and the reader; we obtain the following expression:

$$r = \frac{\lambda}{4\pi}\sqrt{\frac{P_{trans}G_{reader}G_{tag}\left(1-\left|\Gamma_{tag}^{0}\right|^{2}\right)}{P_{th}}} \qquad [2.3]$$

where λ is the wavelength, Γ_{tag}^0 is the reflection coefficient, P_{trans} is the power transmitted, and G_{reader} and G_{tag} are the antenna gain of the reader and the tag, respectively.

However, when communication is taking place over longer distances, whether in the case of active or semi-active UHF RFID, the stresses on the downlink must imperatively be taken into consideration. In this case, for a given chip, the antenna impedance value that maximizes the condition influencing the return link, that is, the ΔRCS or, in other words, the modulation coefficient $K^0 - K^1$, is given by the following formula [BOL 10, CHA 11a]:

$$Z_{a,opt} = R_{a,opt} + jX_{a,opt}$$

with

$$R_{a,opt} = \sqrt{\left(R_c^0 R_c^1 (1 + (X_c^0 - X_c^1)^2 / (R_c^0 + R_c^1)^2)\right)} \qquad [2.4]$$

and

$$X_{a,opt} = -(R_c^0 X_c^1 + R_c^1 X_c^0)/(R_c^0 + R_c^1)$$

In order to validate these theoretical results, we will present an antenna later that maximizes ΔRCS to the detriment of the matching condition. The results of this study are discussed in Chapter 3. Next, we will focus on the practical design of antennas for RFID tags. The very first stage involves the choice and characterization of the chip.

2.3. Discussions about the two chip impedance states Z^i_c

During the design phase of a tag, the design of the antenna necessarily comes after the choice of the chip. For an antenna designer, the chip can be summed up in two values: its input impedance Z_c and its activation power P_{th}. To this, we must obviously add the geometric dimensions and information on the mode of assembling the antenna-chip combination. Note that parasite elements that can be modeled by a series resistance and capacity (R_{ac}, C_{ac})[3] can modify the impedance experienced by the antenna.

However, in reality the problem is more complicated than it appears. Remember the following facts:

– tag communication relies on the switching of chip impedance between two states. This results in two different impedances Z_c^0 and Z_c^1;

3 This information is most often obtained through measurements.

– these impedances are dependent on frequency as well as the power supply to the chip.

Almost certainly, chip manufacturers do not provide information on the non-absorbing state Z_c^1. However, they do give the frequential behavior of Z_c^0, specifying the capacity C_c and the resistance R_c of its equivalent electrical model. It is important to note that not having information on the second state of the chip limits possibilities in terms of design and optimization. In comparison to the two measurable values characterizing the tag's performances (ΔP_{rec} and ΔRCS), in simulations only the activation power can be drawn. With regard to behavior in terms of power, here too, very little information is available. The impedance values are given for a specific power, usually the minimum activation power P_{th}. As we will see in Chapter 3, we can also add that all these parameters are relatively difficult to measure in practice, requiring the setup of specific measurement benches. It is also true that these values generally vary depending on the requests sent to the chip (reading mode: reading, writing, etc.).

Along with information on the chip, it is also necessary to understand the manufacturing procedure to have an idea of the materials used to produce an antenna (the metal and substrate used). We must also take into consideration the chip's report mode on the antenna, which will enable us better to take parasite elements into account (R_{ac}, C_{ac}). In many applications, the choice of these materials is dictated by cost imperatives. If we look at passive UHF tags, they must be mass-produced, and thus at a lower cost. This has resulted in the choice of relatively standardized manufacturing methods. Based on this, the use of very cheap dielectrics (mainly very thin polyethylene terephthalate (PET) plastic materials 50 µm in thickness) is favored. For these same reasons, aluminum is often preferred to copper. This choice, which is based on price rather than on EM characteristics, may have an impact on the final performance of the tag.

Before looking at the antenna design approach for UHF RFID tags strictly speaking, we must examine other elements proper to RFID. The fields of application of RFID are highly varied, and it should be noted that tags are placed on objects of every shape and composition.

In order to facilitate the practical deployment of RFID technology, and again for reasons of price, a tag must usually be able to be used in as many applications as possible (diversity of objects being tracked, heterogeneity of tag densities, diversity of associated reading methods, etc.). Moreover, tags are most often produced to operate on flat supports. The dimensions of the objects to be tracked will dictate those of the tags; there are also several more or less "standard" tag dimensions (90×10 mm^2, 90×30 mm^2, 70×70 mm^2, etc.). Note, however, that in comparison to the wavelength at UHF frequencies (310 mm at 960 MHz in free space), these dimensions are small. This means that RFID antenna designers must employ miniaturization techniques. Passive UHF RFID antennas are mainly flat dipole-type antennas, and the principal miniaturization technique simply consists of folding the arms of the dipole so as to obtain the desired form while retaining the correct EM characteristics.

Not knowing the type of material on which the tag is placed has repercussions on the approach to designing a tag antenna. In reality, designers must design an antenna without knowing the tag's direct environment, even though this environment plays an important role in the tag's performance. The solution is to try to design tags that will be as robust and resistant to their environment as possible. In practice, the classic design approach leads to design tag in free space while seeking to maximize the tag's frequency operating band. Then, characterizations can be made by placing the tags on dielectrics of various permittivities in order to assess their robustness within their near environment. This totally empirical approach is based on the observation that the presence of a dielectric near an antenna has a tendency to move down the operating frequency of the antenna. Additionally, the larger the operating band is in open air, the more it is expected that antenna performance will be high in practice in disrupted environments. As we will see, there is a more evolved solution that takes the presence of a complex environment into account during the design phase of the tag and is used to produce robust tags [CHA 10a, CHA 11b].

Now let us look at the practical application of designing antennas for passive UHF RFID tags. We have seen that the stresses are very high and that there is little maneuvering room. Antenna design consists of finding just the right compromise. We will see that the constraints of miniaturization involve a reduction of antenna bandwidth, and therefore tag performance is necessarily limited.

Operating frequency	Minimum activation power P_{th}	Input impedance Z_c^0	Parallel capacity and resistance	Coupling capacity C_{ac}
840-960 MHz	-15 <-> -19 dBm	$24 - j195$	890 fF / 1.7 kΩ	≈100 fF

Table 2.2. *Approximate values of principal parameters involved in the design of antennas for passive UHF RFID tags*

2.4. Rules of design for RFID antennas: classic design approach

Given all the constraints noted above, we might also think this is a dangerous exercise. It is not, but we might safely say that this specific context is partly responsible for the lack of reliability of UHF technology sometimes observed in practice. This also explains why the design of UHF RFID antennas remains highly empirical and requires a great deal of knowledge.

To sum up what we have just described, the designer begins work with four values as a starting point: R_c, C_c, R_{ac} and C_{ac}; the real and imaginary parts of the chip, and the parasite values connected to the mounting of the chip on the antenna, respectively. Using these four items of data, the designer can access a model describing the frequential behavior of the chip impedance experienced by the antenna for the absorbent state $Z_c^0(f)$.

$$Z_c^0(f) = \mathrm{R}_c + \mathrm{R}_{\mathrm{ac}} + \frac{1}{j\omega(\mathrm{C}_c+\mathrm{C}_{\mathrm{ac}})} \qquad [2.5]$$

In addition, in order to evaluate digitally the performance of the tag produced (most often in terms of read range), the designer uses the value P_{th} provided by the supplier. They will work to design an antenna that is capable of powering the chip on the largest frequency band possible. An example of the values that these parameters may take is shown in Table 2.2. Then, using these parameters, the design of an antenna can begin. The antenna is obviously not an interchangeable component, and will be designed for a given chip, and in some cases for a given application. It is important to keep in mind that chip performance determines the limitations of the tag; given this, the designer's objective is to produce an antenna capable of pushing these limitations as far as possible. With $Z_c^0(f)$ and P_{th}, the designer is able to design an antenna so as to optimize the transfer of power between the reader and the chip, that is, the downlink.

2.4.1. *Classic antenna design approach for passive UHF RFID tags*

Unlike HF RFID antennas, in UHF there are no analytical formulas enabling us to establish the link between the geometry of an antenna and its physical properties (input impedance, gain, etc.). Therefore, an EM simulator must be used. Since the structures are mostly planar, 2.5D simulator tools are most often employed. From there, it becomes a matter of knowing which approach the designer must use to produce the antenna. Above all, the design approach is based on the knowledge and experience of the designer. Here, we will describe an approach with three very distinct stages (see Figure 2.5) [PER 12, RAO 05a].

The first stage consists of sizing a loop around the chip so as to counterbalance its capacitive part. Somewhat similarly to the design of an HF tag, we are seeking to ensure that the loop + chip system resonates around the desired UHF frequency. The other advantage of the presence of the loop lies in the fact that it facilitates near-field communication. The second stage consists of adding metal ribbons, like the dipoles that act as radiative elements. These may be either connected physically to the loop, or positioned near it to create a coupling between the loop and the antenna (Figure 2.5).

The way in which the loop is connected to the metal arms is decisive, as the spacing between the two arms (physical connection) or between the arms and antenna (coupling) are key parameters with a direct impact on the operation of the antenna. While the total length of the antenna can impact the resonating frequency, these spacings make it possible to control the frequency band of the tag. To reduce overload, the lines are bent in twists and turns, such as meanders, or in any way as long as original shapes are produced. To boost bandwidth, rounded shapes are preferred over right angles. A quick optimization inspection phase validates the topology tested, or not.

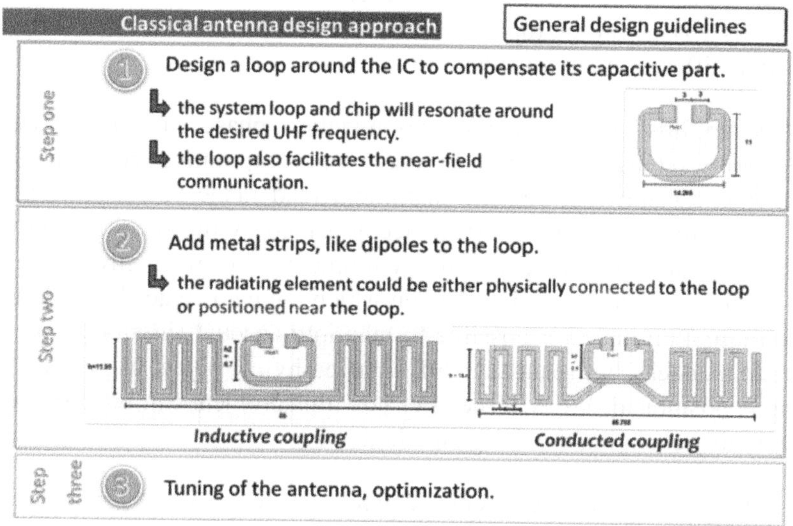

Figure 2.5. *Illustration of the classic antenna design method for passive UHF RFID tags*

The third stage consists of a parametric study focusing on the geometric dimensions of the antenna (spacing between the arms, total antenna length, etc.) and enables an understanding of its electrical behavior (see Figure 2.6). In reality, the challenge of design is to find an antenna topology where a net correspondence exists between the variation of certain geometric dimensions and the variations in the antenna's EM behavior. The designer seeks to identify the key geometric parameters, that is, the parameters that will impact a specific electrical value of the antenna, leaving the others unchanged as much as possible (for example, a dimension that enables us to set the imaginary part of the antenna's impedance, with the real part remaining unchanged). It is possible to represent the results of parametric studies in the form of abacuses (see Figure 2.6), actualizing in this way the link that exists between the geometry of an antenna and its electrical behavior. These abacuses thus characterize the topological behavior of the antenna being studied, showing notably its advantages and limitations. In practice, they also have the major benefit of making it possible to change the dimensions of an antenna quickly (based on the topology studied), but for another chip, that is, with a different input impedance.

From this point, a classic optimization phase can begin. This generally consists of finding the best compromise between the antenna's bandwidth and the maximum read range value. Note that this optimization phase must be accompanied by the specification of the antenna to a given external environment. This given environment may be free space or a dielectric on which the tag is placed. The latter configuration makes it possible to take into account (in an extremely basic manner) the frequency shift induced by the object on which the tag is positioned. We will see later how this problem can be mitigated by totally changing the design method.

When the parasite elements R_{ac} and C_{ac} are not known with great precision, the initial design is readjusted after a prototype is produced. In simulations, the designer focuses on calculating the power reaching the chip P_{chip}, or the reading distance r depending on frequency [2.3]. If we consider that a tag with power P_{chip} is greater than P_{th} functions, we can determine the frequency band of the tag. With regard to the antenna, for purely applicative questions, the objective for tags is to have an isotropic radiation diagram so that they can respond in the same way regardless of their position or orientation in relation to the reader. Gain is not a parameter to be optimized. Only the adaptation level Γ_{tag} contributes to the improvement, or lack thereof, of tag performance. In this case, then in simulation, it is preferable to optimize the Γ_{tag} parameter [2.1] rather than the read range r [2.3]. It is important to note that given the performances of the chips and the emission power enabled, the matching level does not need to be lower than –10 dB, as for classic antennas. A Γ_{tag} on the order of –5 dB is sufficient to guarantee a level of performance in terms of acceptable operating distance (note that Γ_{tag} is not directly the S_{11} parameter of the antenna, but is rather a function of Z_c [2.1]). However, the work of an antenna designer consists of guaranteeing this level of matching over the widest frequency band possible. If we consider the RFID frequency bands authorized in the various regions of the world (see Chapter 1), and if we seek a certain flexibility of use geographically speaking (a tag that can be used anywhere in the world), the tag must be able to function on a frequency band higher than 100 MHz, with tags of small size in relation to wavelength. This significant bandwidth also contributes a certain flexibility of reading compared to a variable environment (compensation for the discrepancy induced by the surrounding dielectrics). This explains why passive UHF RFID antennas often have relatively low levels of adaptation compared to classic antennas.

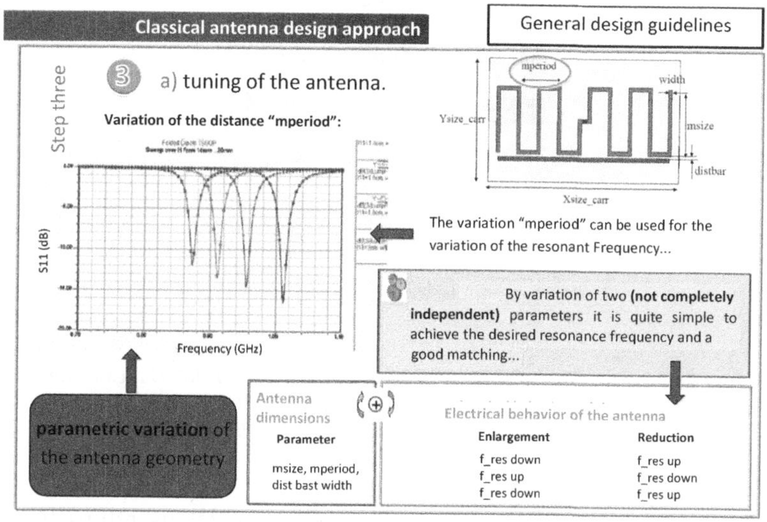

Figure 2.6. *Illustration of the 3rd stage (antenna tuning) of the classic antenna design method (see Figure 2.5). The idea here is to isolate the key geometric parameters while putting parametric studies in place. The results are represented in the form of abacuses, enabling the linking of the antenna's electrical behavior and its geometry*

Everything we have just said can be very easily observed by looking closely at commercial tags. As shown in Figure 2.7, we can see that, whomever the manufacturer, the tag is composed of a loop and two radiative arms, like a folded dipole. If we consider the last three tags manufactured by Avery Dennison (Figure 2.7), they have the particular characteristic of having the same topology while operating with different chips, that is, with different impedances. Complex conjugate matching has been carried out by adjusting geometric parameters such as the total length of the dipole or the thickness of certain parts of the strips. A final optimization limited to very slight variations of an equally limited number of parameters may explain the small differences observed between the three antennas.

Inlay	Tag Manufacturer	Chip
AD-222	Avery Dennison	Impinj Monza 2
AD-224	Avery Dennison	NXP UCODE G2XM
AD-223	Avery Dennison	Impinj Monza 3
ALN-9640	Alien Technology	Alien Higgs 3

Figure 2.7. *Practical examples of commercial passive UHF RFID tags using the design method described here and shown in Figure 2.5. Photo of tags from different manufacturers and information on the chips used*

Before moving on to automated tag design, let us return to the classic method, and more specifically the tuning phase. We can see very clearly that this approach is purely empirical, and that it is a way to avoid the use of analytical formulas used, as in HF, to link the antenna's EM behavior and its geometric dimensions. However, it remains difficult to implement and requires a great deal of experience on the part of the designer. This is why studies have focused on generalizing this approach, by seeking to (1) automate the key parameter search phase and (2) characterize antenna performance in terms of adaptability as exhaustively as possible, that is, the ability to modify certain parameters so that the antenna can function with other chips [LAC 10]. In practice, when a new chip is put on the market, it is important not to have to repeat the whole design process for an antenna, but to start from the previous form and modify it in order to adapt it to the new chip. In this way, this technique makes it possible both to rethink antenna design for RFID tags and to indicate how the design of a known antenna should be modified in order to be able to reuse it with a new chip, that is, with an input impedance different from the one used for the reference tag. One methodology approach is based on the possibility of coupling the database obtained, via a series of parametric studies, with artificial neural networks (ANNs), which will serve as tools for selecting the appropriate geometric configuration. Thus, ANN is capable, using a learning mechanism, of establishing the link between certain physical parameters of an antenna and the EM characteristics of that same antenna. Concretely speaking, as shown in Figure 2.8, a neural network has been obtained with an antenna input impedance (real and imaginary parts) according to frequency and, at output, the values of the corresponding geometric parameters. This approach enables the following developments:

1) it makes it possible to select the most highly adapted geometric parameters, that is, the parameters with a significant impact on the variation of the antenna's input impedance. This approach, then, provides a quantifiable means of reducing the number of these parameters to characterize a given antenna's topology, for a given impedance variation range. If we take an antenna and conduct a parametric study on a large number of its geometric values, using an ANN, we will be able to obtain the corresponding antenna impedance variation range. Based on this, it is possible to repeat this same approach, this time progressively eliminating certain variable geometric parameters while trying to keep the widest possible impedance range;

2) in the end, it is possible to associate an ANN establishing the correspondence between the antenna input impedance and the value of its geometric parameters. This is to some extent a solution to an inverse problem. The geometric modifications of the topology of the antenna being studied are then perfectly characterized and controlled; thus, it is directly possible to make the necessary adjustment of certain antenna parameters in order to satisfy new requirements in specifications without having to change radically the antenna's initial

form. For example, we can use an antenna design with a new chip, or maximize the read range for an RFID frequency sub-band. An example of this type is shown in Figure 2.7.

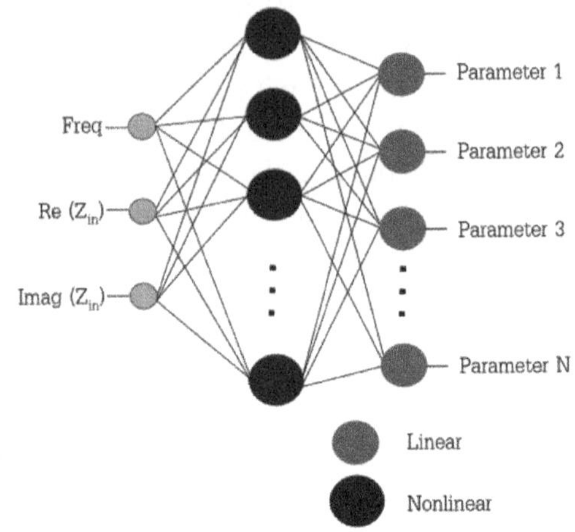

Figure 2.8. *Example of a neural network used to aid the design of RFID tags*

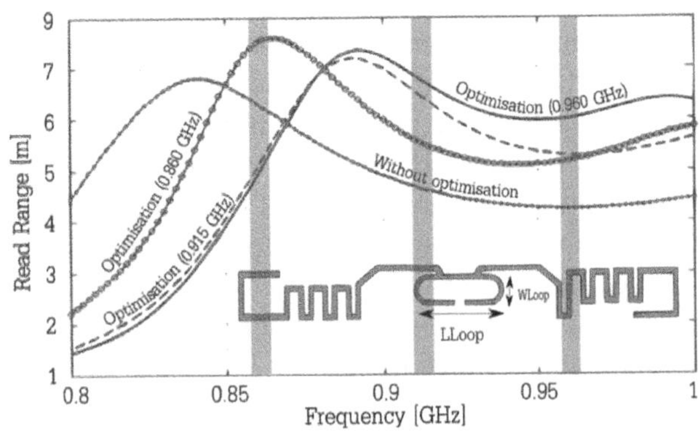

Figure 2.9. *Read range of the antenna of an NXP FFL 95-8 tag, where the LLoop and WLoop have been readjusted based on the neural network characterizing the antenna in a way so as to optimize the antenna's performance for frequencies of 860, 915, and 960 MHz. The chip used here is the Alien Higgs-3, with an impedance of 12-j173 Ω. This impedance is different from the one for which the antenna was designed (the NXP GXL2 chip); this is why the "Without Optimization" curve is not adapted to RFID frequencies. For a color version of this figure, see www.iste.co.uk/perret/radio.zip*

2.5. Robust RFID antenna design methodology

2.5.1. *Context of study*

An innovative design approach that takes into account the immediate environment of a tag during the design phase was introduced in [CHA 10a, CHA 09, CHA 10b, CHA 11b]. This research was carried out as part of a project dealing with the implementation of RFID in the retail supply chain. It constitutes the main part of the PhD done by Chaabane [CHA 12]. The objective of the study principally consisted of designing tags that would be as robust as possible in terms of communication, with the requirement that they would remain compatible with current RFID manufacturing standards. This issue leads, logically, to the examination of tag design methods – as described above – in themselves. The observation is clear; these methods do not allow us to take into account variations in the immediate environment to which a tag may be subjected. This is why a new tag design approach has been introduced. In the end, as we will see, a computer program for the automated design of antennas for passive and robust UHF RFID tags has been developed [PER 10]. In order to remain as general as possible, both dielectric and metallic environments were taken into consideration.

2.5.2. *Description of principle applied*

The approach developed relies on the following principle: nearly all antenna topologies are generated automatically and selected according to the constraints imposed (Figure 2.10). From this, it is possible to break with the empirical approach classically applied, which consists of optimizing a previously used topology. It is interesting to combine EM simulation software with calculation software; the main reason for integrating these two types of software is to take advantage of data manipulation capabilities, the very wide range of functions, and the extended graphic possibilities of software, such as Matlab, while also benefiting from high-performance, multifunction EM simulators. An optimization process based on the concept of genetic algorithms (GAs) is used to satisfy the constraints imposed during the design phase. Optimization relies on an iterative process that (1) generates the shape of an antenna and (2) simulates the antenna and assesses its performances depending on the constraints imposed. Thus, the antenna's shape changes from one iteration to another according to the evolutionary principle used. This is repeated until an antenna is simulated that best fulfills the specifications imposed. An example of antenna topologies obtained in this way is shown in Figure 2.11.

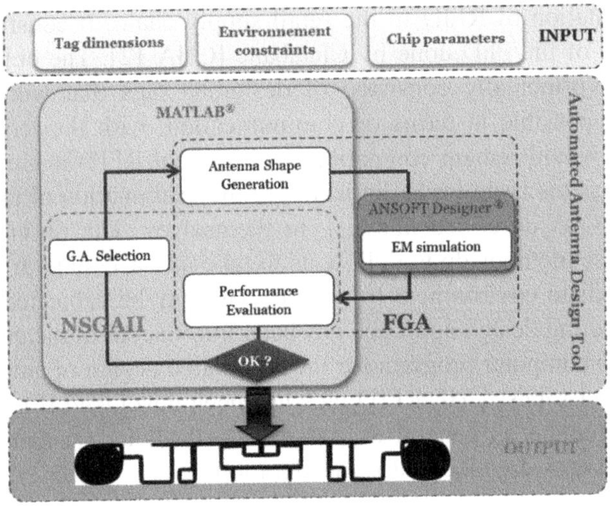

Figure 2.10. *Theoretical schema of automatic antenna design tool for passive UHF RFID tags*

2.5.3. *Principle of co-simulation*

Like most software, Ansoft Designer can be controlled in different ways. The most common way is to use the interactive graphic interface, in which the user manually executes the commands necessary to simulate a given device. A second way of proceeding consists of writing scripts in Visual Basic language that, once executed, generate the same operations as the user could have launched from the graphic interface [LEE 08]. The approach chosen consists of directly inserting the specific functions of Ansoft Designer into Matlab code. This means that everything is controlled from Matlab, which makes this approach quite flexible. In this way, it is possible to avoid manual repetitive tasks both in parameter-setting and in the collection and formatting of results. This is also why the entire process can be automated.

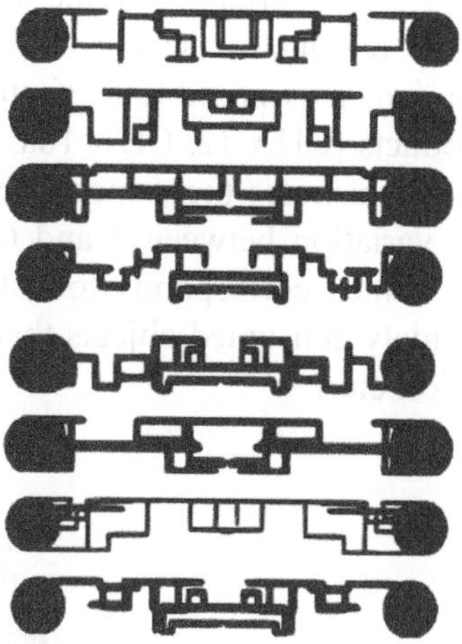

Figure 2.11. *Example of antennas obtained using the developed method of entirely automatic passive UHF RFID label design*

2.5.4. *Taking into account of the environment, design of robust tags*

The universal tag, able to be used independently of the product being tagged and in any environment, does not exist. Therefore, we must still group products into families. The determination of these families is based mostly on the disruptions that these objects might cause to the EM behavior of a tag's antenna. We may distinguish between families of products that are predominantly dielectric, and those that are predominantly metallic, and we will discuss the former scenario here. It has been shown that predominantly metallic objects can also be treated using a similar approach [CHA 12].

For the predominantly dielectric family of products, the permittivity and loss values and geometry (particularly thickness) affect the antenna's behavior. We will see that it is possible to take into account the disruptions induced by a potentially variable operational environment during the design phase. The first stage consists of modeling the disruptive environment to the tag as simply as possible, so as to be able to integrate it into simulations. The idea is not to attempt to simulate a real case precisely (otherwise the result would be specific to this environment); but rather to apply a technique to take into account variations of the dielectric environment in simulation. Thus, by adding to the tag a dielectric plate whose permittivity can be varied, it is possible to simulate the behavior of certain families of products [CHA 12, CHA 10a, CHA 09, CHA 10b, CHA 11b]. Figure 2.12 illustrates the principle used. In the example shown, a permittivity variation between 1 and 6 is considered. This range of permittivity values corresponds to a large amount of the permittivities of the widely consumed objects that surround us (paper, cardboard, some plastics, etc.).

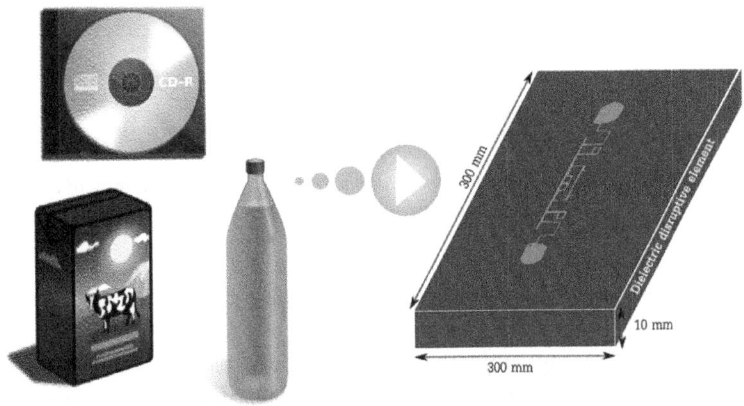

Figure 2.12. *Taking into account of the environment in the automated design process of robust RFID tags. The EM disruption of the tag induced by a family of products is applied in the simulation by adding to the tag a dielectric plate of variable permittivity and a thickness of 10 mm*

Based on this, for each antenna generated, a simulation including the dielectric plate is carried out for each permittivity value retained. In each simulation, performance in comparison to specifications is calculated and is used as input parameters for the iterative antenna selection/generation process. Unlike the "free space" scenario, here it is possible to take tag performance into consideration in relation to the influence of dielectrics, that is, to different environments. The antenna chosen must have the best as possible read range for each frequency point simulated, and for each environment considered.

2.5.5. *Use of the cosimulation principle in the optimization process*

To satisfy the constraints imposed during the design of an antenna for an RFID tag, we use a GA-type iterative process [DEB 02]. GAs have been used successfully in microwaves and for antenna optimization, among other things [HAU 95, JOH 97, LEE 08]. Their use is well suited to problems with a large number of unknowns. The novelty here is that their use is not to optimize a structure, but to design an antenna automatically and completely. We are here taking advantage of the ability to consider a large number of unknowns, as well as to impose multiobjective criteria.

The principle of the automatic design tool is based on an iterative process that consists of generating an antenna form, simulating it and evaluating it in relation to the constraints imposed. The generation of forms from one iteration to another is carried out on the basis of evolutionist principle proper to GAs (see Figure 2.13). This is repeated until an antenna design is obtained that best satisfies the specifications. To design antennas that are robust and environment-resistant, antennas that have a wide read range and whose sensitivity to the environment is as low as possible (that is their sensitivity to the presence of the dielectric with different permittivities; see Figure 2.12) are sought. Moreover, this behavior is desired for the entire RFID frequency band (840–960 MHz). In order to translate the constraints we have just specified into language that can be understood by the design tool used, the concept of cost function is used (also called objective function or fitness function). These functions will link the EM properties of antennas simulated on Ansoft on the one hand and the GA implemented on Matlab on the other hand. In other words, they make it possible to quantify the level of sufficiency of the performances of each antenna with regard to the constraints imposed. Insofar as the AG used [DEB 02] is multiobjective, as many functions of this type as desired can be defined, with the additional possibility of weighting these functions so as to impose a kind of hierarchy.

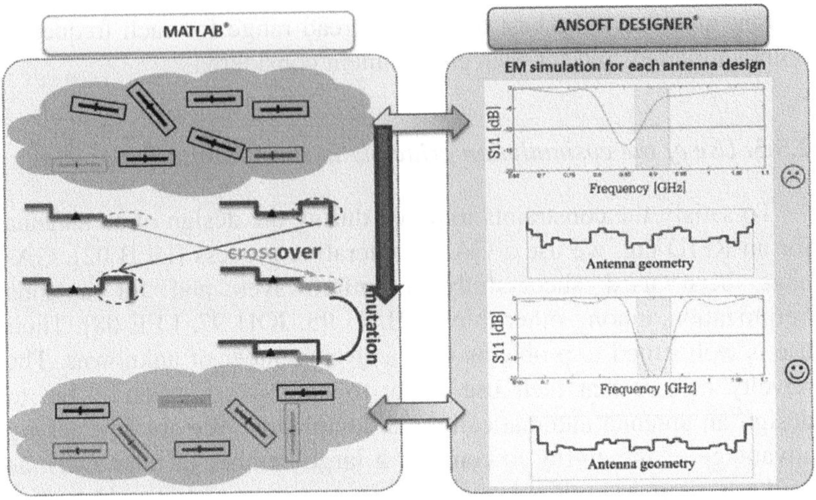

Figure 2.13. *Schematic illustration of the evolutionist character of genetic algorithms. The initial population includes a certain number of antennas. The simulator is then used to obtain the EM characteristics of each individual member of the initial population. Next, Matlab evaluates the performances of each member with regard to the criteria imposed using fitness functions that make it possible in the end to compare the potential of individual members. For a color version of this figure, see http://www.iste.co.uk/perret/radio.zip*

As seen above, read range is given by equation [2.3]. It is expressed in terms of various parameters: frequency, threshold power P_{th}, maximum effective radiated power (ERP) emitted by the reader $P_{trans}G_{reader}$, tag antenna gain G_{tag}, and adaptation coefficient Γ_{tag} between the antenna and the chip. This formula is based on the radar equation, and is therefore valid only for far-field activity.

Using equation [2.3], we can define the fitness function as follows:

$$F_{cost} = \begin{bmatrix} \min_{i=1:3}\left(\alpha_i r(f_i, \varepsilon_{r\,1})\right) \\ \min_{i=1:3}\left(\alpha_i r(f_i, \varepsilon_{r\,2})\right) \\ \min_{i=1:3}\left(\alpha_i r(f_i, \varepsilon_{r\,3})\right) \\ \cdots \\ \min_{i=1:3}\left(\alpha_i r(f_i, \varepsilon_{r\,n})\right) \end{bmatrix} \qquad [2.6]$$

where i represents the index related to the frequencies f_i chosen (e.g. $f_1 = 868$ MHz, $f_2 = 915$ MHz, $f_3 = 955$ MHz), α_i is a weighting coefficient on the frequencies, $r(f_i, \varepsilon_{r\,n})$ is the read range, which is in this case a function of frequency as well as of the permittivity $\varepsilon_{r\,n}$ of the *n*th dielectric considered in the model. The other read range parameters are set in the following manner:

– $P_{trans}G_{reader} = 2$ W (maximum transmission power authorized by European regulations);

– P_{th} = -14 dBm (value considered for chip NXP GXL2 [NXP]);

– G_{tag} and Γ_{tag} are obtained via simulation on Ansoft Designer.

It should be noted that the fitness function F_{cost} returns n real values, which describe for each dielectric the minimal performance on the frequency band being studied. In the GA used, the iterative process seeks to maximize each cost-function value. This is why, in the end, the tag obtained will be the one with the largest read range, taking into account various disruptions in the simulated environment. We can see via the fitness function expression [2.6] how it is possible after the design phase to introduce the concept of tag sensitivity in relation to its operating environment. In the next section, we will discuss how it is possible to generate antenna forms automatically.

2.5.6. *Generation of antenna forms*

Each iteration of the process includes the generation of an antenna form, its simulation and a fitness function evaluation. In this design tool, the idea is to generate antenna forms that are as varied as possible, all with a limited number of unknowns (in the order of a few dozen). It is of interest to be able to reproduce common forms of RFID antennas in addition to generating original forms. Moreover, the manner of generating antennas presented here is based on the folded dipole principle: the form of the antenna is obtained by putting conductor wires end-to-end (in planar technology, this corresponds to rectangles) so as to have fairly ordinary shapes. This approach makes it possible to produce antennas with reduced dimensions, as well as the ability to create loops. This last point is important because, as we have seen, loops serve in RFID both to adapt a chip to an antenna and to enable the tag to function in close-field activity.

Here, we will take the example of a symmetrical wire antenna. To generate the antenna, it is sufficient to define the total number of rectangles *N*, their width L_w, and the two parameters defining the position of each contiguous rectangle, one in relation to the other (see Figure 2.14). These parameters will be optimized throughout the process. This configuration enables us to generate various forms from one simple, elementary shape: the rectangle. It also makes possible for good flexibility in the generation of the antenna's shape, while requiring a limited number of unknowns.

As shown in Figure 2.14, we write L_i to represent the length of the *i*th rectangle, L_w for the width of the set of rectangles, and θi for the angle formed by the arms i and i+1. The placement of the chip is potentially variable (in the example, it is variable only vertically). Once the first arm of the antenna is generated by this procedure, the second arm is obtained by symmetry. However, it is possible to add minor variations between the arms in order to make the shape asymmetrical. Finally, to generate antennas, it is enough to define the list of values for the set of parameters describing the structure of the antenna, specifically:

– N: the total number of elementary rectangles;

– L_W: the width of all the rectangles;

– C_p: the vertical position of the chip;

– $\begin{pmatrix} L_i \in [L_{min}, L_{max}] \\ \theta_i \in [0, \frac{\pi}{2}, \pi, \frac{3\pi}{2}] \end{pmatrix}_{i=1:N}$: the length of each rectangle as well as the angles formed by the rectangles with each other.

To build each wire, we will consider the two end points, that is, points A and B defined by the following expression:

$$[A_{(x_i,y_i)}, B_{(x_{i+1},y_{i+1})}]: \begin{pmatrix} x_{i+1} = x_i + L_i * \cos(\theta_i) \\ y_{i+1} = y_i + L_i * \sin(\theta_i) \end{pmatrix}$$

These parameters are the unknowns of the problem and will be adjusted during the antenna design process. Thus, this approach enables us to generate diverse forms from simple combinations of elementary ones. This form of antenna also calls for a limited number of unknowns (2N +3).

A final aspect concerns the maximum dimension authorized for the tag (written as L_t and W_t). The antenna is generated so that it occupies an area previously determined by the designer. Tests focusing on the modification of this area are also conducted [CHA 11b]. We can see in Figure 2.14(c) some antennas obtained using this approach for areas of 80 × 80 mm², 40 × 40 mm² and 95 × 8 mm².

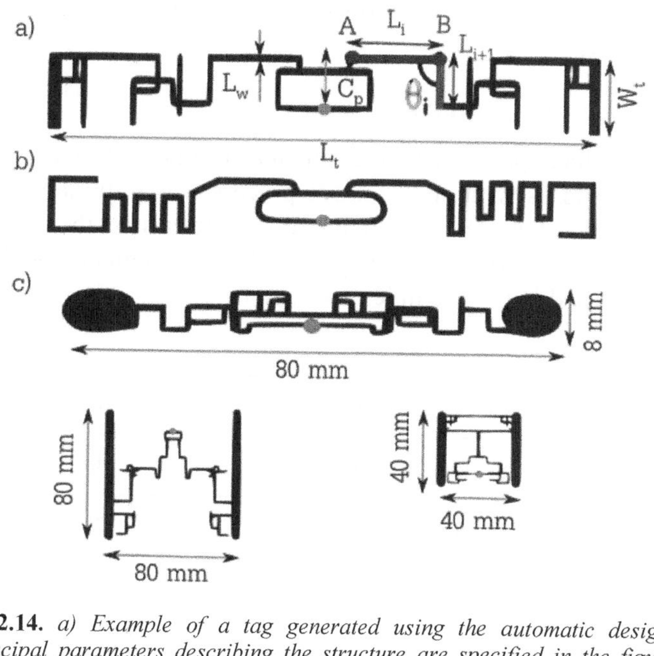

Figure 2.14. *a) Example of a tag generated using the automatic design tool. The principal parameters describing the structure are specified in the figure:* C_p*: vertical position of the chip;* L_w*: width of the wire;* L_i *and* L_{i+1}*: successive lengths of rectangles;* L_t *and* W_t*: maximum dimensions allowed. b) Reference tag: NXP (FFL95-8). c) Topology of antennas obtained for three different dimensions: Tag 1: 80 × 80 mm²; Tag 2: 40 × 40 mm²; Tag 3: 95 × 8 mm². The antennas are not all represented on the same scale. All of the tags shown operate with the same chip*

2.5.7. *Application of the automated design tool via an example*

2.5.7.1. *Manufacturing constraints*

For the design and production of tags, we consider the standard constraints in RFID tag manufacturing. Here, the tag must have a maximum size of 95 × 10 mm^2. The antenna is produced using aluminum with a thickness of 9 µm, printed on PET with a relative permittivity of 3.2 and a thickness of 50 µm. The chip used is the GX2L by NXP, with a complex input impedance Z_c^0= 22-j195 Ω at 915 MHz. In this first example, the objective is to obtain an antenna with a maximal read range in free space. The manufacturer NXP sells the tag FLE 95-8 operating with chip GX2L. This tag is shown in Figure 2.14(b); it will henceforth be considered as the reference tag for this study.

2.5.7.2. *Convergence of antenna form*

As explained previously, the antenna design tool is an iterative procedure, and as with any iterative process it is important to study convergence. Likewise, it is of interest to determine the simulation time associated with each design. Table 2.3 summarizes the AG parameters used in the example described here.

Total dimensions of antenna (L_t and W_t)	95*10 mm^2
Number of chromosomes (population)	40
Number of iterations (generation)	28
Number of elementary rectangles	120
Paramter variation range (mm)	Li=[2.8] Lw=[0.5, 1.5] Cp=[0.5, 5]

Table 2.3. *Typical genetic algorithm parameters*

The approach proposed makes it possible to obtain a result satisfying technical specifications after an execution time of around 10 h (EM simulation of an antenna takes around 1 min) with a Dual Core processor (2 GHz, Memory: 4 GB). Convergence is typically obtained after 30 generations. While significant, this simulation time does not require any human interaction and the design process is totally automated. It can also be slightly reduced by using a more high-performance computer.

If we look more closely at the antenna obtained, which is shown in Figure 2.14(a), we can see that the approach automatically converges toward an antenna with a loop. As shown in Figure 2.11, which presents a panel of antennas obtained with the automatic design tool, this is not isolated behavior. It is linked to the fact that we are seeking here to optimize the read range in relation to the expression of the uplink [2.3], that is, the complex conjugate matching impedance. We will see in the next chapter that when we attempt to design antennas with other criteria (for example, for more exotic applications) the antennas obtained no longer include loops [CHA 11a]. It is of interest to study the evolution in antenna forms throughout the iterative design process. This allows us to retrace the evolution of the form as well as the appearance of the loop. Figure 2.15 shows this evolution according to generation, specifying the fitness function associated with the antennas. The antennas presented are those with the highest fitness function within the population being considered. We can see in Figure 2.15 that an extremely wide panel of different antenna topologies is obtained throughout the iterative process. This is a highly favorable point, as it ensures the creation of an antenna fulfilling specifications with a high success rate. We can also see that a loop appears in the 10th generation, and that after that, the antennas demonstrating the best performance always include a loop. In the end, the antenna retained after 28 generations (see Figure 2.14(a)) shows a form that is fairly similar to the NXP reference antenna (see Figure 2.14(b)), even if only in terms of the dimensions of the loop, with regard to all of the antenna forms appearing throughout the design process.

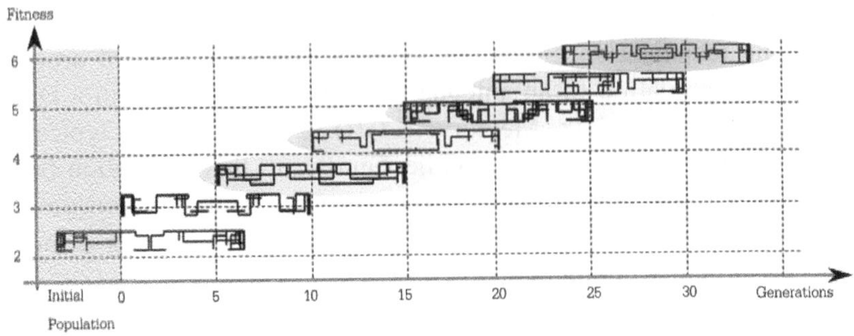

Figure 2.15. *Evolution of the shape of antennas generated, tested and selected throughout the process of RFID tag design. Only some of the antennas generated are shown. For a selected generation, we are showing the antenna with the best fitness function. The more the number of generations (abscissas) increases, the more the fitness function improves (ordinates)*

2.5.7.3. *Results obtained*

To demonstrate the interest of this tool, we began by applying it to a hypothetical case in which the antenna is planned to operate in free space. In this case, there is no need to model the external environment, since it is considered to be both known and fixed. Figure 2.16 shows an antenna obtained using the design tool under these conditions. This antenna is then placed on a dielectric plate whose permittivity is then caused to vary. In this way, we are presenting the read range [2.3] for each permittivity value over the whole RFID band being considered. We can clearly see the negative impact of the dielectric plate on the antenna's performance; the higher the permittivity, the smaller the read range. We can also see that, for a permittivity value of 5.7, the read range can fall below 2 m for certain frequencies. This behavior is identical to that observed when the classic antenna design method is used. In fact, if we take the NXP antenna (see Figure 2.14(b)), it shows behavior that is similar in every respect [CHA 11b]. Since the antenna has been optimized in free space, it has a very good read range in this case (greater than 9 m), but any variation in the environment will negatively influence its behavior to a significant extent.

Now, let us look at an antenna obtained with the automatic design tool, this time taking environmental influence into consideration. The antenna thus produced, as well as its read range according to frequency and for multiple dielectric values, is shown in Figure 2.17. Here, we can clearly see that its behavior is stable whatever the dielectric considered; in fact, the read range is permanently situated at between 5 and 7 m. We are, therefore, in a scenario where the external environment will have much less impact on the antenna performance. Measurements with everyday objects have been taken and confirm these results [CHA 11b]. Note also that this robust character has a detrimental effect on the maximum performance of the tag when the latter is placed in free space.

The experimental validation was carried out in an anechoic chamber in monostatic mode [POU 10]. As shown in Figure 2.18, the measurement bench is composed of a radio frequency (RF) vector signal generator (Agilent MXG-N5182A) used for the transmission part as an RFID reader emulator. A spectrum analyzer (Tektronix RSA3408A) is used as a receiver. We measure the minimum power to be sent to the RF vector signal generator in order to enable activation of the tag (the tag is considered to operate when we see on the spectrum analyzer that the whole RFID frame is sent by the tag). This process is repeated for all of the frequencies considered. Figure 2.19 shows a comparison between the simulation and the measurement in a scenario in which the tag is in free space and in another scenario in which the tag is placed on a dielectric plate with a permittivity of 2.2 and a thickness of 1 cm.

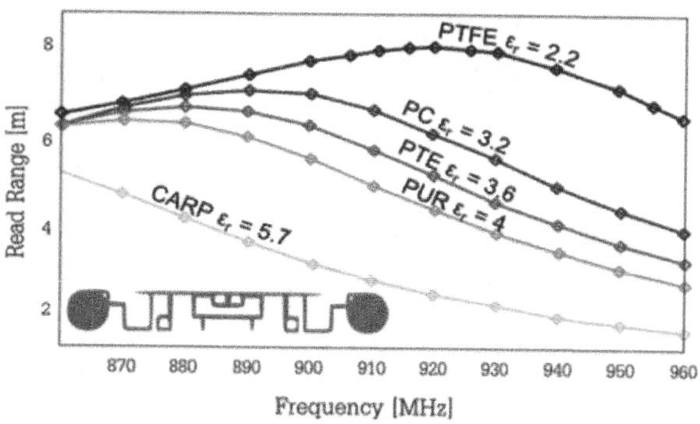

Figure 2.16. *Simulated read range [2.3] versus the frequency for an antenna created by the automatic design approach to operate in free space. Once the antenna is created (shown at the bottom of the figure), it is simulated in the presence of a dielectric plate (Figure 2.21) whose permittivity is varied between 2.2 and 5.7*

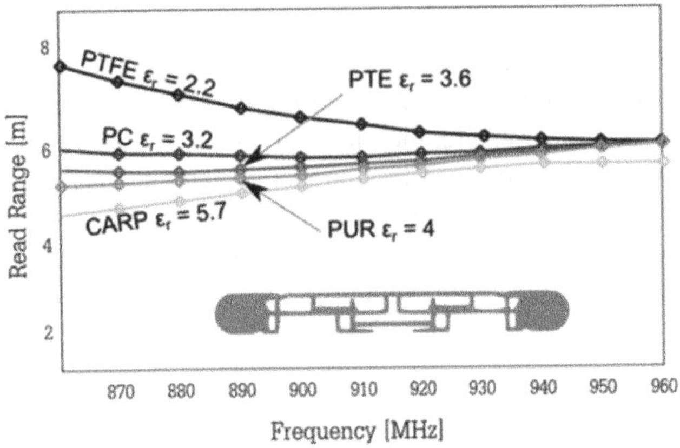

Figure 2.17. *Simulated read range [2.3] versus the frequency for an antenna produced using the automatic design approach to function on a family of low-permittivity objects (from 2.2 to 5.8). Once the antenna is created (shown at the bottom of the figure), it is simulated in the presence of a dielectric plate (Figure 2.21) whose permittivity is varied between 2.2 and 5.7*

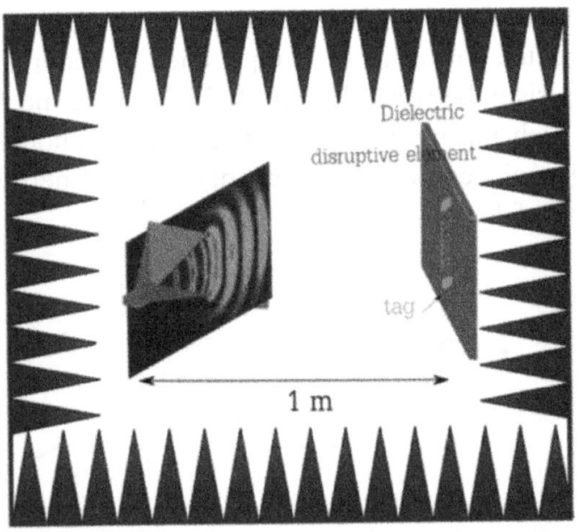

Figure 2.18. *Measurement bench applied to characterize RFID tags. We measure precisely the minimum power to be sent to cause the tag to function. For a color version of this figure, see http://www.iste.co.uk/perret/radio.zip*

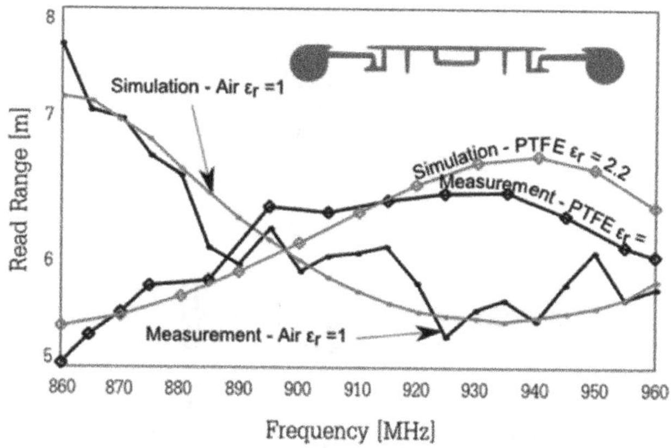

Figure 2.19. *Comparison of measurements – simulation of read range versus the frequency for an antenna produced using the automatic design approach. Two configurations are shown: the free space scenario and the scenario in which the antenna is positioned on a dielectric plate with a permittivity of 2.2 and a thickness of 1 cm*

2.5.7.4. *Metal objects*

The design procedure has been tested in several other configurations, notably for the creation of tags operating near metallic materials. This configuration is particularly difficult to process, since metal materials will impact tag performance more significantly than dielectric materials. In addition, the idea here is always to preserve compatibility with low-cost RFID production processes, that is, to stay with devices printed on a single layer. However, a method using the design tool has been successfully applied and validated. The model used in Ansoft Designer to take metallic materials into account is shown in Figure 2.20. The results show that metallic geometry must be taken into account in a more detailed way than dielectric geometry. Thus, to design a tag that will operate on a metal casing (Figure 2.20(b)), a model based on the use of three metal plates (Figure 2.20(c)) has had to be used. The method of generating antenna forms that has proved to perform most effectively is very similar to that used for dielectric configurations – specifically, the placing of end-to-end metal rectangles.

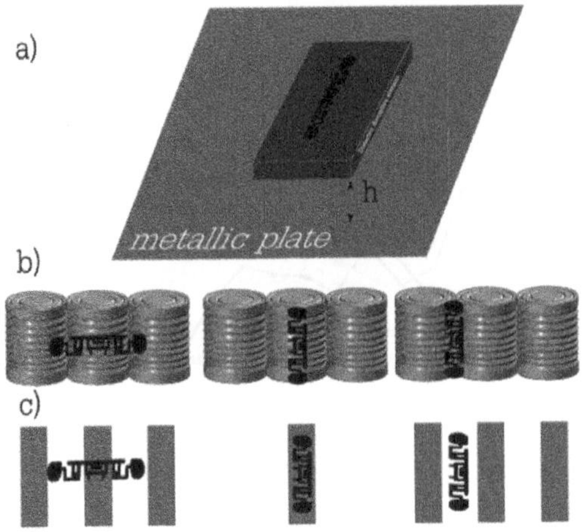

Figure 2.20. *Model used in Ansoft Designer for the taking into account of a metallic environment. a) Simple model based on the use of a metal plate. b) Real metallic environment studied (here of the metal "can" type – h=1–2 mm). c) Model used to simulate case study b)*

2.6. Conclusion

The heavy constraints currently imposed on RFID have a direct impact on the antenna design methods used until now. It is clear that the antenna design approach for passive UHF RFID tags is a specific one, for various reasons. We may cite the adaptation that takes place for loads that are not purely resistive, while playing solely on the antenna's geometry. There is also the lack of knowledge of the support (or supports) on which the antenna will be used. From there, design approaches must make it possible to produce tags that will be resistant to various environments. Finally, economic constraints impose both small dimensions and a wide operating frequency band. Thus, the levels of adaptation sought are lower than for classic antennas.

A great deal of research in the field of tag design has been conducted, aimed most often at moving toward a "universal" tag which, given current chip performances, is still a goal rather than a reality. The development of automated design approaches that take as many constraints as possible into account appears vital for the eventual success of this research.

We have introduced the classic approach to passive UHF RFID tag design. Like antenna design in general, an empirical approach is used systematically to create tags. An experienced tag designer will begin with an antenna form predetermined from experience and then optimize it. The antenna topologies that will be tested in this way are necessarily limited in number; their potential in terms of performance is stopped once the form is chosen. In addition, it is a very delicate matter to be able to impose a large number of external constraints on the antenna (taking into account of the environment, robust antennas, etc.).

Next, we have introduced an original antenna design methodology for creating RFID tags that are robust to the environment. The automated design approach applied in the RFID domain has made it possible to obtain antennas with original forms that satisfy strong constraints. The flexibility of antenna forms has considerably increased the level of freedom of structure, which allows us in the end to address problems with extremely demanding specifications. This approach is based on the principle of the joint use of different complementary commercial software. Original antenna topologies are generated automatically and selected according to the constraints imposed. The prototype thus obtained shows comparable behavior, whether in contact or not with a dielectric object that may disrupt it. Moreover, the approach used is totally automatic, and thus significantly reduces the man–machine interaction time required, shortening the design phase as a result.

2.7. Bibliography

[BOL 10] BOLOMEY J.C., CAPDEVILA S., JOFRE L., *et al.*, "Electromagnetic modeling of RFID-modulated scattering mechanism. Application to tag performance evaluation", *Proceedings of the IEEE*, vol. 98, no. 9, pp. 1555–1569, 2010.

[CHA 09] CHAABANE H., PERRET E., TEDJINI S., "Conception automatisée d'un tag RFID UHF robuste", *16èmes Journées Nationales Micro-Ondes*, Grenoble, France, 2009.

[CHA 10a] CHAABANE H., PERRET E., TEDJINI S., "Towards design of robust UHF RFID tag", *4th IEEE RFID Conference*, Florida, pp. 223–229, 14–16 April 2010.

[CHA 10b] CHAABANE H., PERRET E., TEDJINI S., "Computer aided-design for robust UHF RFID antenna", *Wireless Systems International Meeting WSIM'10*, Campina Grande, Brazil, 2010.

[CHA 11a] CHAABANE H., PERRET E., DAIKI M., *et al.*, "RFID tag design approach based on chip multi states impedance", *Asia-Pacific Microwave Conference Proceedings (APMC)*, Melbourne, pp. 1466–1469, 2011.

[CHA 11b] CHAABANE H., PERRET E., TEDJINI S., "A methodology for the design of frequency and environment robust UHF RFID tags", *IEEE Transactions on Antennas and Propagation*, vol. 59, no. 9, pp. 3436–3441, 2011.

[CHA 12] CHAABANE H., Contribution à la conception d'antennes pour tag RFID en bande UHFl, PhD Thesis, University of Grenoble, 2012.

[DEB 02] DEB K., PRATAP A., AGARWAL S., *et al.*, "A fast and elitist multiobjective genetic algorithm: NSGA-II", *IEEE Transactions on Evolutionary Computation*, vol. 6, no. 2, pp. 182–197, 2002.

[FIN 10] FINKENZELLER K., *RFID Handbook: Fundamentals and Applications in Contactless Smart Cards, Radio Frequency Identification and Near-Field Communication*, Wiley, 2010.

[GRE 63] GREEN R.B., "The general theory of antenna scattering", OSU Report No. 1223-17, 1963.

[HAR 63] HARRINGTON R., "Electromagnetic scattering by antennas", *IEEE Transactions on Antennas and Propagation*, vol. 11, no. 5, pp. 595–596, 1963.

[HAU 95] HAUPT R.L., "An introduction to genetic algorithms for electromagnetics", *IEEE Antennas and Propagation Magazine*, vol. 37, no. 2, pp. 7–15, 1995.

[JOH 97] JOHNSON J.M., RAHMAT-SAMII Y., "Genetic algorithms in engineering electromagnetics", *IEEE Antennas and Propagation Magazine*, vol. 39, no. 4, pp. 7–21, 1997.

[LAC 10] LACOUTH P., PERRET E., FONTGALLAND G., *et al.*, "New RFID's chip matching technique using artificial neural networks", *2010 IEEE International Conference on RFID-Technology and Applications (RFID-TA)*, Guangzhou, China, 2010.

[LEE 08] LEE K., KIM Y., CHUNG Y.C., "Design automation of UHF RFID tag antenna design using a genetic algorithm linked to MWS CST", *4th IEEE International Symposium on Electronic Design, Test and Applications (DELTA '08)*, Hong Kong, 2008.

[NIK 05] NIKITIN P.V., RAO K.V.S., LAM S.F., *et al.*, "Power reflection coefficient analysis for complex impedances in RFID tag design", *IEEE Transactions on Microwave Theory and Techniques*, vol. 53, no. 9, pp. 2721–2725, 2005.

[NIK 06] NIKITIN P.V., RAO K.V.S., "Theory and measurement of backscattering from RFID tags", *IEEE Antennas and Propagation Magazine,* vol. 48, no. 6, pp. 212–218, 2006.

[NXP 08] NXP Application Note, AN 1629 UHF RFID Label Antenna Design, UHF Antenna Design, September 2008.

[PAR 09] PARET D., *RFID at Ultra and Super High Frequencies: Theory and Application*, Wiley, 2009.

[PER 10] PERRET E., CHAABANE H., TEDJINI S., "RFIDTool", Dépôt du logiciel auprès de L'Agence pour la Protection des Programmes (APP) Inter Deposit Digital Number: IDDN.FR.001.390041.000S.C.2010. 000.20600, October 2010.

[PER 12] PERRET E., TEDJINI S., NAIR R., "Design of antennas for UHF RFID tags", *Proceedings of the IEEE,* vol. 100, no. 7, pp. 2330–2340, 2012.

[POU 10] POUZIN A., VUONG T., TEDJINI S., *et al.*, "Bench test for measurement of differential RCS of UHF RFID tags", *Electronics Letters*, vol. 46, no. 8, pp. 590–592, 2010.

[RAO 05a] RAO K.V.S., NIKITIN P.V., LAM S.F., "Antenna design for UHF RFID tags: a review and a practical application", *IEEE Transactions on Antennas and Propagation*, vol. 53, no. 12, pp. 3870–3876, 2005.

[RAO 05b] RAO K.V.S., NIKITIN P.V., LAM S.F., "Antenna design for UHF RFID tags: a review and a practical application", *IEEE Transactions on Antennas and Propagation*, vol. 53, no. 42, pp. 3870–3876, 2005.

[SKA 09] SKALI S., CHANTEPY C., TEDJINI S., "On the measurement of the delta radar cross section (ΔRCS) for UHF tags," *IEEE International Conference on RFID*, p. 34, 2009.

3

New Developments in UHF RFID

3.1. Introduction

The ultra high frequency radio frequency identification (UHF RFID) sector is in constant evolution due to its economic importance. Today, we are observing a trend aimed at developing new applications and functionalities based directly on existing RFID systems. One of the strengths of RFID lies in the fact that this technology is (1) extremely widespread, (2) mature and (3) regulated by clearly established norms. Based on this, the temptation is strong to attempt to reuse this technology so as to add functionalities to it, but without modifying it in depth. Thus, RFID can be seen as a system of communication in which we are attempting to combine its current identification function with other functions. These other functions – and more specifically the ones that will be presented below – all go back more or less to the concept of wireless sensors. As we will see, it is possible to seek to render RFID tags sensitive to a physical value, in such a way that this value can be measured. It should be noted that this information, which is intrinsically linked to the physical aspect of the tag, is not directly part of the RFID communication protocol, but is rather superimposed in a way on the classic signals. Thus, it requires specific processing by the reader in order to be accessible to the user.

Two applications, which are different enough to be distinguished between, will now be introduced. The first application is a method of wireless measurement of the input impedance of RFID antennas and is based on the knowledge of the chip's two impedance states. The second application is a very general approach aimed at transforming an existing tag into a sensor that can be used to monitor environmental parameters in real time. Thus, it is possible to measure the level of a liquid in a container, the distance between a tag and a metal plane, or temperatures and humidity levels. We will see the full interest of this approach, in that no major modifications would need to be made to the RFID system itself. The result would be a sensor function associated with a unique identifier (ID), all relying on a communication system that is already widespread. This fundamental advantage of RFID lies in the self-powering of tags and wireless communication, which makes the exchange of information much quicker and more efficient. Ultimately, it is possible in this way to obtain low-cost sensors with long lifespans.

3.2. Wireless measurement technique for antenna impedance

We will see in this section how it is possible, using a classic RFID system, to measure the impedance of an antenna without wires. On the one hand, the principle is based on the use of an RFID chip whose characteristics, in this case the two impedance states, must be precisely known. An interrogation request from the tag will make it possible to activate the chip and to cause it to switch between these two separate impedance states. On the other hand, the approach is based on the application of a measurement technique called far-field reflectometry, which is used to evaluate the performances of electrically small antennas. This method has the advantage, among others, of eliminating the influence of the environment near the antenna, for example, the cable that links it to the measurement device during characterization in an anechoic chamber.

If we take the example of UHF RFID, the precision of impedance measurements plays a crucial role in the design of RFID antennas for tags. Most UHF RFID tag antennas are balanced structures [HAI 10, MEY 98], such as dipoles or loops. These antennas must be powered by differential input signals with no real ground plane. Therefore, the impedance of a balanced antenna cannot be measured directly by most instruments, which terminate in asymmetrical ports such as coaxial ports. When a symmetrical antenna is connected to an asymmetrical measurement port, the currents powering the two arms radiating from the antenna are unequal, and consequently the antenna's impedance cannot be measured with precision. This is why the antenna's input impedance cannot be characterized easily and precisely using the two single-ended ports of a vectorial network analyzer [LEO 07, MEY 98]. Since most vector network analyzers (VNAs) are equipped with asymmetrical ports, special procedures must be used to characterize a symmetrical device. Moreover, the test bench used in the antenna measurement process also causes other parasite effects, which must be taken into account in order to obtain a precise measurement.

There are multiple methods that can be used to measure the input impedance of a balanced device from single-ended ports. The following approaches can be used; they (and their principal limitations) are listed below:

– An external balun. The balun is used to force the circulation of equal currents in each of the radiating arms of the antenna. In this case, the precision of the measurement is dependent on the precision of the balun itself [HAI 10, LEO 07].

– A large ground plane using antenna symmetry and image theory. The precision of the measurement is dependent on (1) the size of the ground plane compared to the antenna, (2) the connection between the measurement cable and antenna [LEO 07] and (3) a virtual ground linking two coaxial cables with their external connector while also connecting them to the two differential ports of the VNA on the one side and to the antenna on the other side. From there, we can see that it is possible to calculate the input impedance from the S parameters [QIN 09]. The precision of the measurement relies on the transition added to link the devices to one another. It also requires significant postprocessing to reach the value desired.

This is why the use of a measurement method without cabling is of particular interest, as it resolves many of the difficulties listed above, particularly those related to connectors. Thus, the method that we will discuss here, and which consists of the principle of relying on the RFID system itself to extend its functionalities, will be generally applicable to the measurement of electrically small antennas and, more particularly, to the measurement of RFID tag antennas. However, in order to apply this principle of far-field reflectometry coupled with the RFID approach, the first task is to procure the values of the two impedance states. To do this, it is vital to have a reliable measurement bench. This is why, before describing the method of measuring antenna impedances, we will begin by describing how to characterize a functioning RFID chip.

3.2.1. *Characterization of RFID chips and measurement of the two impedance states*

In order to use RFID tags for sensor-type applications or, more precisely, to measure the impedance of an antenna, we must have exact knowledge of the value of the chip's two impedance states. This information, as we saw in Chapter 2, can often be accessed only via measurements. This is why we will now introduce a complete characterization method for RFID chips.

3.2.1.1. *Introduction*

As we have seen, an RFID tag is composed of an RFID chip soldered to the terminals of an antenna. The principle of communication relies on simple modulation principles. The RFID norm sets the frame of the reader to interrogate the tag. The answer is also normalized and is based on a retro-modulation of the signal transmitted [NIK 06]. To establish this communication, and to retro-modulate the signal received, the chip must present two impedance states, Z_c^0 and Z_c^1. As with every electronic component, the electrical characteristics of the chip vary according to frequency and power supply. The impedance values of the chip are thus a function of frequency f and power P and can be written as follows: $Z_c^i(f, P), \mathrm{i} = 0,1$.

To use these tags as sensors, it is important to be able to characterize the two impedance values of the chip precisely according to frequency and power. As we saw in the last chapter, the manufacturer assigns only one impedance value, for a single power level. Frequency dependence is generally available for a given power, the threshold power P_{th}. It is provided by an equivalent electrical circuit (an equivalent resistance and capacity, as shown by expression [2.5] in Chapter 2). Thus, only one measurement can be used to access all of the chip's characteristics.

Several approaches and measurement benches have been developed in the literature [AND 14, CHE 09, MAY 08]. These measurement benches are based most often on the use of specific instruments, such as probes or tuners, which can disrupt the measurements with their effects, and remain fairly complex methods to implement. Moreover, these approaches do not always enable access to the behaviors of the two impedance states depending on power and frequency. The method briefly presented here is based on the idea of taking the measurement when the chip is in communication. From there, we can access all of the data desired [DAI 11].

3.2.1.2. *Description of measurement bench*

The difficulty of this sort of measurement lies mainly in the fact that information sensing must occur throughout the whole communication phase between the tag and reader. In addition, the measurement bench is composed of an arbitrary waveform generator in order to emulate the request from the reader, a spectrum analyzer to detect the tag's response, and a radio frequency (RF) circulator to switch between transmission and reception, as indicated in Figure 3.1.

To take the measurement, RFID chips are soldered on to a dedicated microstrip line. A Short-Open-Load calibration (SOL) is then used to extract the chip's impedance values.

The results thus obtained are shown in Table 3.1. In Figure 3.2, we can see the real and imaginary parts of the input impedance of chip NXP GX2L according to the power transmitted by the generator for a frequency of 915 MHz. Note that there is a characteristic power level from which the impedance variation changes; this is the threshold power P_{th}. The impedance jump observed corresponds to the activation of the chip, which, therefore, has enough energy to retro-modulate its return message. We can then observe the two impedance states Z_c^1 and Z_c^0.

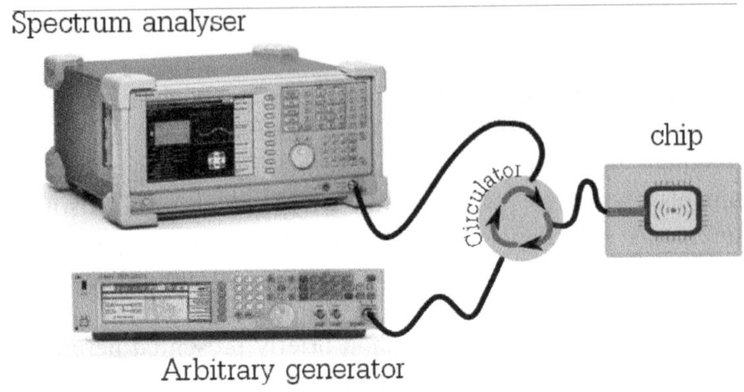

Figure 3.1. *Measurement bench for RFID chip impedances*

Impedance @ 915 MHz in Ω	Manufacturer's datasheet	Measurement
$Re(Z_c^1)$	-	86.4
$Re(Z_c^0)$	22	25.6
$Im(Z_c^1)$	-	-54.7
$Im(Z_c^0)$	-195	-147.7

Table 3.1. *Impedance values measured for the two states of chip NXP GX2L compared to the ones given in the manufacturer's datasheet [NXP 08]*

The variation of the input impedance of the two states of the chip according to frequency for a power level higher than the threshold power P_{th} is shown in Figure 3.3. Thus, we can completely describe the behavior of the chip. This information will subsequently serve as input data. Once the chip is connected to the antenna being tested, it will be used to go back to the impedance of the latter. As we are most often seeking to characterize an antenna's frequency, the interest of this approach is clear; it will allow us to access all of the data that are vital for the next part of the study.

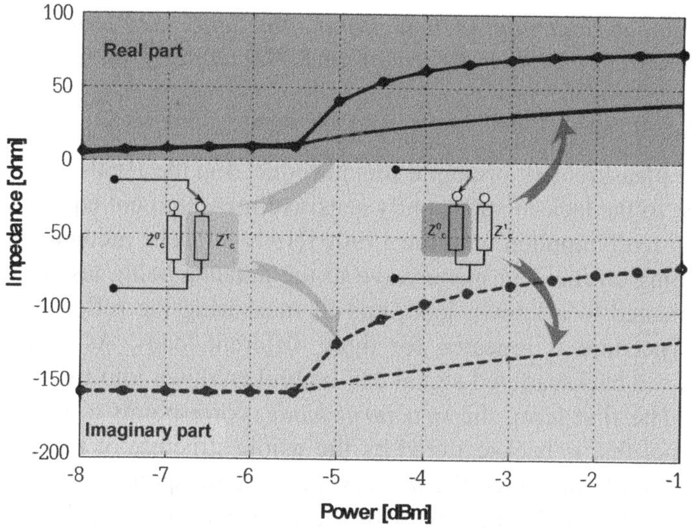

Figure 3.2. *Real impedance (solid line) and imaginary impedance (dashed line) of chip NXP GX2L for the high state Z_c^1 (•) and low state Z_c^0 (no marker) depending on the power supplied by the generator for 915 MHz*

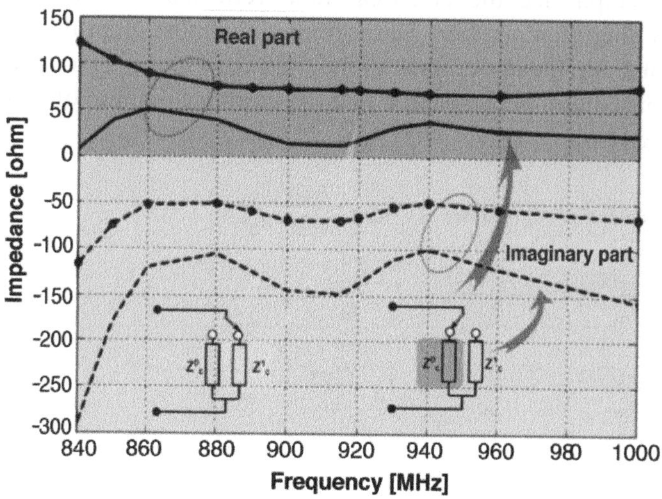

Figure 3.3. *Real impedance (solid line) and imaginary impedance (dashed line) of chip NXP for the high state Z_c^1 (•) and low state Z_c^0 (no marker) depending on frequency for a generator power of –2.6 dBm*

3.2.2. *Theoretical approach to input impedance extraction from a small antenna based on the use of an RFID chip*

3.2.2.1. *Far-field reflectometry measurement technique*

The measurement principle was introduced in the 1960s [HAR 63], but due to the lack of sufficiently sensitive measurement equipment, it was not really applied until the 1990s [HAN 89]. The method consists of sending an incident plane wave to the antenna being tested, which is connected to a known load, and of measuring the reflected wave. This procedure is repeated for three different loads. As Figure 3.4 shows, the backscattered signal can be broken down into two separate terms. The first term, the *structural mode*, corresponds to the part of the signal that is backscattered by the whole structure of the antenna. The second term, called the *antenna mode*, corresponds to the part of the signal that will interfere with the load connected to the antenna [HAR 64].

The complete theory is presented in [HAN 89]. It is interesting to note that several formulations are possible, depending on whether we wish to emphasize the electrical field reflected for one particular charge rather than another. The formulations most often used are the ones that cause the appearance of the field corresponding to a load that is in short circuit [3.1], adapted [3.2], or with a condition of conjugate matching [3.3]:

$$E_{scat}(Z_c) = E_{short} + E_{ant}\frac{I_{short}Z_c}{I_a(Z_a+Z_c)} \quad [3.1]$$

$$\frac{E_{scat}(Z_c)}{I_m^*} = \frac{E_{scat}(Z_a^*)}{I_m^*} + \Gamma^*\frac{E_{ant}}{I_a} \quad [3.2]$$

$$\frac{E_{scat}(Z_c)}{I_m} = \frac{E_{scat}(Z_a)}{I_m} + \Gamma_a\frac{E_{ant}}{I_a} \quad [3.3]$$

with

$$\Gamma_a = \frac{Z_a-Z_c}{Z_a+Z_c} \text{ and } \Gamma^* = \frac{Z_a^*-Z_c}{Z_a+Z_c}$$

where Z_a and Z_c are the antenna impedance and the impedance of any load connected to the antenna, respectively. I_{short} corresponds to the current circulating in the antenna terminals when the latter is short-circuited[1]. Likewise, I_m (and I_m^*, respectively) are the currents in the event that a matched load (or a complex-conjugate matched load, respectively) is placed on the antenna. I_a is the amplitude of the current applied to the antenna that produces a far field E_{ant}. Note that, in compliance with the notations, we have $E_{short} = E_{scat}(0)$.

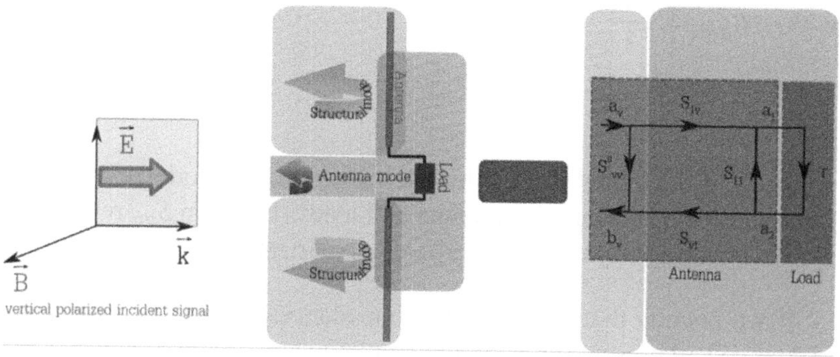

Figure 3.4. *Model of an antenna in terms of structural mode and antenna mode for excitation in copolarization*

Likewise, it is possible to write the expression of the backscattered field showing the field reflected by the antenna in open circuit. We obtain the following expression:

$$E_{scat}(Z_c) = E_{open} - E_{ant}\frac{V_{open}Z_a}{V_a(Z_a+Z_c)} \qquad [3.4]$$

V_{open} corresponds to the tension at the antenna terminals when the latter is in open circuit.

Another formulation leading to the complex radar cross-section (RCS) expressions ($\underline{\sigma}$) of the loaded antenna depending on its parametric characteristics, such as its gain (S_{1v} and S_{v1}) and reflection

1 This makes it possible to set the power level sent to the antenna, which then appears in the equation as the term I_{short}.

coefficient (S_{11}) is given in [HAR 64, WIE 98]. These equations have been used and simplified in [BOR 10] by considering a single polarization. Based on the signal flow graph in Figure 3.4, the relationship between the complex RCS and the various parameters of the model is given by equation [3.5].

$$\underline{\sigma} = \left(S_{vv}^{S} + \frac{r}{1-S_{11}} S_{1v}S_{v1}\right)^2 \quad [3.5]$$

with, $G_{ant} = \frac{\sqrt{4\pi}}{\lambda} \cdot |S_{1v}S_{v1}|$

$$r = \frac{Z_c - Z_0}{Z_c + Z_0}, S_{11} = \frac{Z_a - Z_0}{Z_a + Z_0} \quad [3.6]$$

where Z_0 is the impedance of the line which, in this model, establishes the connection between the antenna and load[2]. The term S_{vv} represents the structural backscatterer in vertical copolarization.

If we assume that we know the impedances Z_a and Z_c, equations [3.1]–[3.5] all present three complex unknowns:

– we have $E_{scat}(Z_c), E_{open}$ and $(E_{ant}\frac{V_{open}}{V_a})$ for [3.4];

– S_{vv}^{S}, S_{11} and $(S_{1v}S_{v1})$ for [3.5].

This is why it is necessary to take three field measurements (amplitude and phase) for three different load values Z_c^i. Thus, we obtain a system with six equations and six unknowns.

Note here that if we are seeking a simple analytical solution for this system of equations, one of the three loads is imposed by the choice of the expression to be used. In [3.1], one of the three loads must be a short circuit, while in [3.3] and [3.5] a matched load condition should be prioritized[3]. This point will be vital subsequently, as in order to

2 Note that in this chapter **r** designates a reflection coefficient (to use the notations of [BOR 10]) and not the read range as in Chapter 2.

3 This is a way of noting that expressions [3.3] and [3.5] are exactly the same. The differences arise from the choice of values used (RCS in one case, electrical field in other case), as well as the taking into account in [3.5] of a feed line (which can be analytically removed by taking $Z_0 = Z_a$).

apply it to RFID, we will see that the short-circuit condition must be favored, and therefore expression [3.4].

A solution of these equation systems with one, two or three load(s) imposed is given in [BOR 10, MAY 94, PUR 08]. The solutions given in [MAY 94] and [BOR 10] are not of great interest for what will follow, since they lead to a scenario that is too highly constrained. For example, in [BOR 10], using formulation [3.5] and the following loads:

– short circuit (CC): $Z_c^{CC} = 0$, which corresponds to $r = -1$;

– open circuit (CO): $Z_c^{CO} = \infty$, which corresponds to $r = 1$;

– adapted load (L): $Z_c^L = Z_a^*$, which corresponds to $r = 0$;

we get:

$$S_{vv}^S = \sqrt{\sigma_L} \qquad [3.7]$$

$$S_{1v}S_{v1} = 2 \cdot \frac{(\sqrt{\sigma_{CO}} - \sqrt{\sigma_L})(\sqrt{\sigma_L} - \sqrt{\sigma_{CC}})}{(\sqrt{\sigma_{CO}} - \sqrt{\sigma_{CC}})} \qquad [3.8]$$

$$S_{11} = \frac{S_{1v}S_{v1}}{\sqrt{\sigma_L} - \sqrt{\sigma_{CC}}} - 1 \qquad [3.9]$$

with σ_{CC}, σ_{CO} and σ_L corresponding to the complex RCS of the antenna connected to a short circuit, an open circuit and a load Z_a^*, respectively.

Note that another, simpler formulation establishing a direct link between the antenna impedance Z_a and the three RCSs can also be deduced from expressions [3.7]–[3.9]. Thus, we have expression [3.10]:

$$Z_a = -\frac{\sqrt{\sigma_{CO}} - \sqrt{\sigma_L}}{\sqrt{\sigma_{CC}} - \sqrt{\sigma_L}} Z_c^L \qquad [3.10]$$

It is possible to directly establish the link between relationship [3.10] and the one given in [MAY 94] [3.11]. The latter is obtained from a formulation similar to [3.1] and assuming an open circuit, a short circuit and any real load. It arises more exactly from the current/tension matrix procedure used by Harrington to address the case of any loaded objects [HAR 64, HAR 96].

$$Y_a = -\frac{E_{CC} - E_L}{E_{CO} - E_L} \cdot Y_c^L \qquad [3.11]$$

with $Y_a = \frac{1}{Z_a}$ the admittance of the antenna being tested, and Y_c^L the inverse of Z_c^L.

Finally, a more general approach given in [PUR 08] gives the solution to the problem obtained from a relationship similar to [3.4], imposing only the open-circuit case. To be precise, the article by Pursula *et al.* [PUR 08] considers the following cases: open circuit, any capacitive load and matched load. However, in reality, the expression given is more general, and only the open-circuit condition is truly imposed. The relationship thus obtained can be expressed as a function of fields, which gives expressions [3.12] and [3.13].

$$Z_a = \frac{Z_c^1 - Z_c^2 A}{A-1} \qquad [3.12]$$

with

$$A = \frac{E_{scat}(Z_c^2) - E_{open}}{E_{scat}(Z_c^1) - E_{open}} \qquad [3.13]$$

where Z_c^1 and Z_c^2 are ordinary loads.

Starting from [3.1], that is, imposing one of the three loads as being a short circuit, we demonstrate that we obtain the following expression in an analogous manner:

$$Z_a = \frac{Z_c^1 Z_c^2 (B-1)}{Z_c^2 B - Z_c^1} \qquad [3.14]$$

with

$$B = \frac{E_{scat}(Z_c^2) - E_{short}}{E_{scat}(Z_c^1) - E_{short}} \qquad [3.15]$$

The same thing can also be done using relationship [3.2] (or [3.3], respectively); this results in other expressions, in which a complex conjugate matching impedance (or a matching impedance) is imposed. We can clearly see here that these expressions are of no interest for the determination of the antenna's impedance. Unlike [3.12] and [3.14], it would be necessary to know the value of the impedance we are attempting to measure[4]. Likewise, we would need to be able in practice to impose this impedance value, which is not constant in frequency, on the whole frequency band being studied. Short-circuit and open-circuit conditions, however, do not pose any particular difficulty.

Note, however, that [MAY 94] and [PUR 08] present a study of the impact of errors made (for example, of measurement) on the extracted antenna impedance value Z_α. It appears from this research that if we impose an open circuit and a short circuit, the third load to be prioritized must be the closest to the matched load scenario.

A validation via simulation of what has just been mentioned is shown in Figures 3.5 and 3.6. Figure 3.5 shows the antenna being studied, which is an RFID antenna, and therefore has a complex input impedance of $Z_a = 22 + j195$ at 915 MHz. The results of simulations obtained on CST MWS ® are shown in Figure 3.6. Impedance extraction based on relationship [3.11] has been carried out for the four load configurations described below (see Figure 3.5(a)):

– Case 1: $Z_c^1 = 0$, $Z_c^2 = \infty$, $Z_c^3 = 15\ \Omega$.

– Case 2: $Z_c^1 = 0$, $Z_c^2 = \infty$, $Z_c^3 = {}^1/_{jCw} = -j193$ with C = 900 fF.

– Case 3: $Z_c^1 = 0$, $Z_c^2 = \infty$, $Z_c^3 = R + {}^1/_{jCw} = 15 - j193$ with R = 15 Ω and C = 900 fF.

– Case 4: the short circuit has been replaced by a capacity (C) and the matched load is formed by a capacity (C) and a resistance (R)

4 It would be necessary to measure $E_{scat}(Z_a^*)$ and $E_{scat}(Z_a)$, and therefore to know Z_a in order to place this load at the antenna input.

connected in a series: $Z_c^1 = {}^1/_{jCw} = -j193$, $Z_c^2 = \infty$, $Z_c^3 = R + {}^1/_{jCw} = 15 - j193$ with R = 15 Ω and C = 900 fF.

Thus, the matching condition $Z_c^L = Z_a^*$ is approached for cases 3 and 4 only. The expected result is also shown in Figure 3.6. The variation of antenna impedance (both the real and imaginary parts) according to frequency can actually be simulated directly on the software by positioning a discrete port at the terminals of the antenna. The other results arise from a plane-wave excitation, with the antenna connected to a discrete element with the desired load. The electrical field is calculated using a probe placed in the far-field zone. We can clearly see in this study that the more strongly the adaptation condition is verified (cases 3 and 4, and particularly around 915 MHz, which corresponds to the adaptation frequency), the better the results will be. Thus, we obtain a high level of correspondence between the impedance extracted using the reflectometry method and the one simulated directly with a discrete port, all within the frequency band of 850–1,200 MHz.

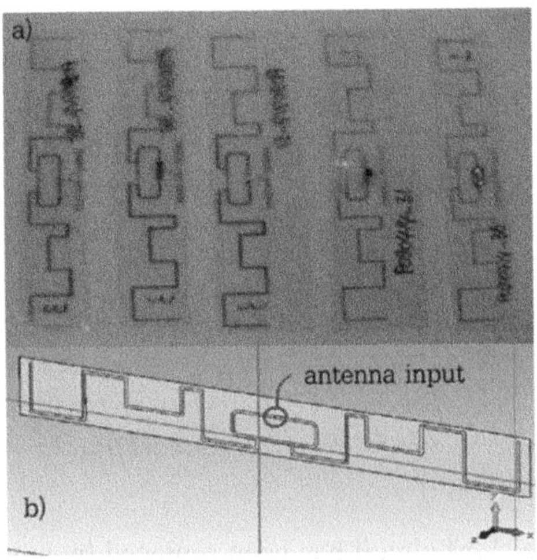

Figure 3.5. *a) Photo of RFID antennas used for measurements. Right to left: tag with NXP chip; tag in short circuit; tag in open circuit; tag with a resistance of 15 Ω; tag with a capacity of 900 fF. b) Its model under CST MWS. Antenna characteristics: length of 100 mm, width of 12 mm, on PET substrate ($\varepsilon_r = 3.2$, 50 µm thick). This antenna was designed to operate with an NXP UCODE G2XL chip with an impedance of $Z_C = 22 - j195$ at 915 MHz*

We can conclude from this study that the error observed is all the weaker as the matched load condition is approached. Compliance with this condition limits the frequency band of validity, as well as the application of the approach itself. For application in RFID, however, we will see that this is not really a problem. Practically speaking, this method can easily be applied to RFID; in fact, in the case of a tag, this matching condition is produced with the low state Z_c^0 of the chip. Open-circuit and short-circuit conditions are also simple to obtain, for example, after having removed the chip from the tag. Therefore, it is quite possible to use this method to extract the impedance value of any small antenna in the UHF RFID band.

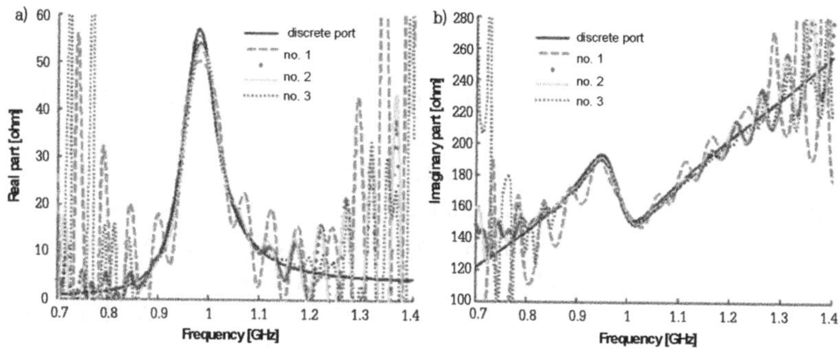

Figure 3.6. *a) Real part and b) imaginary part of the RFID antenna impedance shown in Figure 3.5. Results were obtained using the principle of reflectometry [3.11]. Four different configurations are shown. The "discrete port" curve corresponds to the values sought; that is to the simulation of the antenna's entry impedance carried out with the port connected directly to the antenna's terminals. For a color version of this figure, see www.iste.co.uk/perret/radio.zip*

3.2.2.2. *Principle of reflectometry applied to RFID for the measurement of antenna impedance*

We have just seen that expressions [3.12] and [3.14] can be used to calculate the value of an antenna's impedance if it is also possible to measure the electrical field (amplitude and phase) backscattered by the antenna when the latter is loaded with three different impedance values. In this case, of the three, only one value is imposed; specifically, an open circuit for [3.12] and a short circuit for [3.14].

The question to be asked now is how, in practice, can we easily obtain a device that will switch between two impedance values? The device must be small in dimension so as not to disrupt the measurement. Likewise, in order to automate it and thus gain slightly in precision, the device should be able to be remotely controlled.

It so happens that an elegant and particularly pertinent solution to this difficulty is to use RFID, that is, both an RFID chip and the communication protocol that goes with it. In this case, the two impedance states will be obtained by an RFID chip (see Figure 3.7). This is why the values of the two impedances depending on frequency (and at least for the power used) must be known. The approach described above also enables us to answer this question precisely. Likewise, to cause the two impedance values to switch, it is enough to rely on the RFID communication protocol, for example, by sending an identification request. The switch command is, therefore, given remotely and without a cable. The dimensions of RFID chips (on the order of one square millimeter) respond perfectly to the issue of measurement. With regard to the third measurement (short circuit or open circuit), though it is relatively simple to obtain in practice, it nevertheless requires interaction with the antenna during the measurement. Another limiting factor involves the frequency band. In this case, it is dependent on the operating frequency band of the tag, specifically between 860 and 960 MHz. However, as we will see, it is possible to contribute answers to these questions.

If we return to the third measurement, which cannot be taken with the chip – examining cases of electrically small antennas – we can eliminate the need for it. If we consider this perfectly acceptable simplified hypothesis, especially for RFID tag antennas, the problem can be reduced to a system of two complex equations and four unknowns.

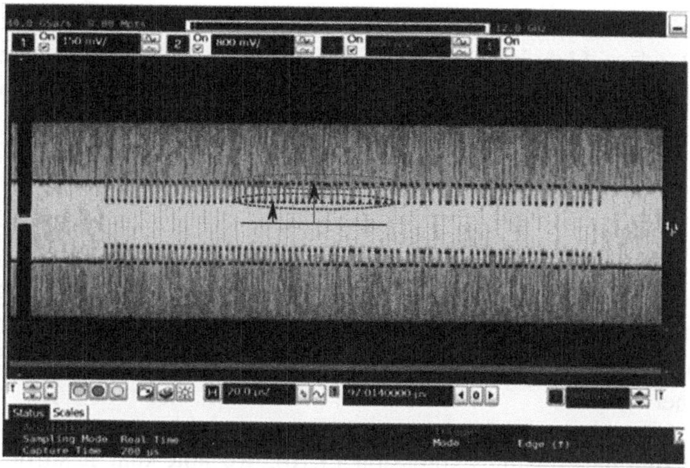

Figure 3.7. *Capture via oscilloscope of a signal reflected by an RFID tag. The two states (high and low) are clearly identifiable on the frame, making it possible to go back to the values* $E_{scat}(Z_c^1)$, $E_{scat}(Z_c^2)$ *(in amplitude and phase). From there, it is possible to obtain the antenna's input impedance using expression [3.16]. For a color version of this figure, see www.iste.co.uk/perret/radio.zip*

For this, we can use expressions [3.12] and [3.13] alone, which take the open-circuit case into consideration. The field reflected when the antenna is in open circuit must be weak in comparison to the field obtained with the loads Z_c^1 and Z_c^2, which leads us to expression [3.16]. This condition is generally true for electrically small antennas; the best example is the half-wave dipole antenna where E_{open} can be considered as zero at the resonance frequency of the dipole.

$$Z_a = \frac{Z_c^1 E_{scat}(Z_c^1) - Z_c^2 E_{scat}(Z_c^2)}{E_{scat}(Z_c^2) - E_{scat}(Z_c^1)} \quad [3.16]$$

If we use Green's expression [GRE 63], which is normally expressed as a function of the antenna characteristics [3.17], the approximation produced corresponds to taking parameter C as equal to 1.

$$\sigma = \frac{\lambda^2}{4\pi} G^2(\theta, \varphi) \left| C + \Gamma^* \right|^2 \quad [3.17]$$

Thus, we obtain the classic expression [3.18] discussed in [HAR 64], which is used to express the RCS of a loaded antenna, simply as a function of the load and antenna characteristics; specifically its gain and input impedance.

$$\sigma = \frac{\lambda^2}{\pi} G^2(\theta, \varphi) \left| \frac{Re(Z_a)}{Z_a + Z_c} \right|^2 \qquad [3.18]$$

It is a simple matter to show that based on [3.18], assuming (1) complex values and (2) two ordinary loads, we get [3.16]. We can see through this simplification how different expressions can come together.

In summary, using expression [3.16], we note that it suffices to connect two loads to the antenna being characterized in order to deduce its impedance (averaging the measurement of the two fields in amplitude and phase, which can also be expressed in complex RCS). These two loads may very well be the two impedance states of an RFID chip, based on the time after which they are known. Thus, using an RFID chip that we cause to switch simply by entering into communication with the tag, it is possible to extract the antenna's input impedance remotely.

This approach is particularly indicated for RFID antennas, insofar as approximation is possible. In reality, miniature or "electrically small" antennas are characterized by small dimensions in comparison to the wavelength, which is the case – as we saw earlier – for RFID antennas. This is all the more interesting because these antennas are hard to characterize in practice, mainly due to the difficulty of connecting them to a cable (which will then disrupt the antenna). As we have seen, they are usually sensitive to the surrounding metallic and dielectric materials.

In terms of limitation, it is important to note that the approximation described above is applicable only with difficulty to antennas with a loop at the chip. In this case, the antenna in open circuit will behave in a manner similar to a short-circuited (dipole) antenna, that is, with a high RCS. As we noted earlier, the presence of a loop is fairly classic in RFID, used to compensate for the imaginary part of the chip and thus to create a condition of conjugated adaptation, which enables us to maximize the distance across which the tag can respond. We will see that when we focus on the downlink as well [2.4]; that is, when we attempt to obtain the optimal antenna impedance value, the topologies of the antennas created by the automated design tool no longer include a loop. This result leads us to think that, in this case, since the weight on the complex conjugate matching condition is less, antennas without loops may appear. In every case, whether the antenna includes a loop or not, formula [3.12] remains valid and will be applicable whatever the form of the antenna.

With regard to the frequency range of this approach, standard tags are made to operate within the UHF band, that is, between 860 and 960 MHz. However, we can see in practice that for a simple operation (such as a request for a tag's authentication), it is possible to make the chip switch over a much wider range of frequencies. In fact, the 800 MHz–1.1 GHz band can be reached without great difficulty. In addition, if we wish to access other frequency bands, it is always possible to turn to other families of RFID chips. As an example, the UCODE HSL IC – SL3ICS3001 chip from NXP is a dual-frequency chip, functioning on the 2.4–2.5 GHz band as well as the UHF band [NXP 12]. Likewise, the RFID chip from TOPPAN FORMS appears to be able to function over the whole 13.56 MHz–2.45 GHz band.

This first example gives a general idea of the interest that may lie in using an RFID system to take measurements. Next, we will look at other examples that demonstrate the significant potential of RFID for creating wireless sensors.

3.3. Toward the use of RFID as a sensor

In the previous section, we saw the difficulty of designing tags able to function independently of the immediate environment for which they are used. In practice, anyone using an RFID system has noted the impact of the presence of a metal object on its read range, for example. There are numerous examples in this area. Figure 2.16 of Chapter 2 shows to what extent the antenna designed is sensitive to its dielectric environment, and more generally to the physical phenomena surrounding it. In practice, we can clearly see the effects on communication of humidity or the proximity of metallic or dielectric materials, and even the coupling effect observed when several tags are close to one another. For applications of identification and traceability, these effects are quite harmful, and the objective is, as we saw in Chapter 2, to reduce them as much as possible. Thus, we opt for robust antennas, like the one shown in Figure 2.17 of Chapter 2. In this case, antenna performance (here in terms of read range) is as stable as possible in comparison to the immediate environment of the tag, but and also to frequency. However, what is a disadvantage in the classic use of RFID becomes an undeniable advantage if we are seeking to add a sensor function to that of identification. This will be possible to the detriment of read range. Thus, classic antennas will respond directly to this new difficulty. We will generally opt for the antennas shown in Figure 2.16 or, better still, for tags which are on the commercial market, and therefore directly operational, like the NXP tag, etc.

In the case of the use of tags as sensors, the principle is to exploit the tags' sensitivity to their environment. If we are able to link the disruption of tag performance to the variation of an environmental value (movement, temperature, pressure, etc.), we can then quantify this variation and operate a standard tag as a sensor. This idea has been presented in several articles [BHA 10a, MAR 08], where it has been shown that it is possible to use the RFID communication protocol, a standardized and mature protocol, to execute a wireless sensor function.

Thus, the possibilities offered by the use of standard RFID tags as sensors are considerable, and still little-exploited. Available information about the electrical characteristics of the chips that are at the heart of the use of tags as sensors is, as we have seen, most often incomplete or even missing. It is important, therefore, to be able to characterize all of the parameters of the RFID chip; specifically, its two impedance states and its variation depending on frequency and input power. In reality, the chip's nonlinear behavior will be added to the variations we wish to measure, which complicates the extraction of the value we are seeking to measure [AND 14]. Lack of knowledge about the chip only complicates the approach.

From there, turning our attention either to the variation of the minimum power needed to activate the tag (downlink), or to the variation of power received by the reader (uplink), it has been shown that it is possible to produce wireless, low-cost sensors of temperature [BHA 11, MAN 12], humidity [SID 07], movement [BHA 09] or liquid level [BHA 10b] while retaining the identification function as well. Thus, we can obtain low-cost sensors with long lifespans, as long as the tag is a relatively simple component (we will see later how it is possible with chipless technology to produce components that are far simpler).

However, it is important to note the principal limitations of this approach. To produce a sensor, it is not enough to measure a signal, which will move depending on the value we are seeking to acquire. It is above all necessary to ensure that this modification arises exclusively from the value in question. This is a matter of selectivity; the variation of other environmental parameters must not significantly modify the signal being measured. As we have just stated, in RFID a large number of external factors have an impact on the backscattered signal. However, we are operating under the notion that this functionality must not result in a more expensive system for the user, but should be freely available. After that, it is up to the user to use it if it can respond to these issues. This will be able to be the case for "controlled" environments, that is, where there are most often few variables in time, where only one parameter (the one being measured) will have a real impact on the tag.

For proof of the concept, we can look analytically at Figures 3.8, 3.9 and 3.10. The power values brought into play (notably P_{chip} and ΔP_{rec}) are nonlinear functions of the physical values we are attempting to monitor. All of this relies on the idea that all of the RFID's characteristic values are functions of the antenna impedance of tag Z_a. This impedance is itself also dependent on the antenna's environment, and particularly on the variations it may undergo as time passes. Generally speaking, disruptions can be grouped into two types: modifications of the immediate metallic environment, and those of the electrical environment, which can be modeled as a variation of the antenna's surrounding effective permittivity. In this way, a large number of sensors can be produced. Thus, it is possible to detect the level of a liquid directly, provided only that the tag is affixed to the (non-metallic) container whose water level we are seeking to measure, for example. The same is true for measuring humidity, as this directly modifies the relative permittivity of the tag's antenna. With regard to temperature, we can act cleverly by updating tag-sensors with a mobile metal part confined within an environment that will change from the solid state to the liquid state depending on temperature, in order to use it as a threshold sensor [BHA 11]. It is also possible more simply to use a material whose permittivity is a function of temperature [MAN 12]. Thus, we can see that it is possible, after a calibration phase, to go back to the physical values we are attempting to measure. Figures 3.9 and 3.10 schematically and analytically present the principles applied to correlate the measurable physical values to the characteristic values of RFID. The RFID values in question are the ones introduced in Chapter 2.

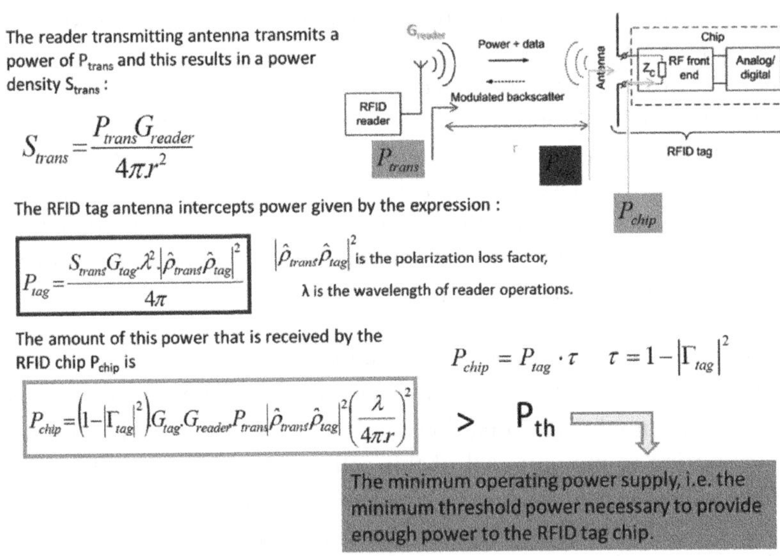

Figure 3.8. *Power balance applied to an RFID system. Thus, we obtain the expression of power reaching the chip P_{chip} depending on various parameters such as the coefficient Γ_{tag}. If this power P_{chip} is greater than the activation power of the chip P_{th}, the chip has enough energy to function*

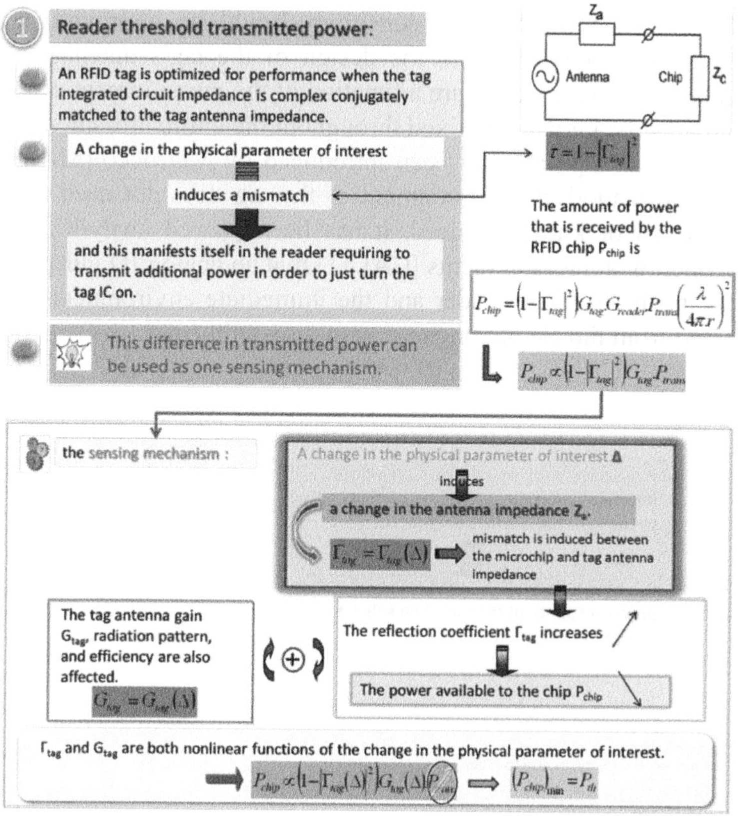

Figure 3.9. *Operating principle of a sensor based on RFID technology. In this first configuration, the information sought is given by the minimal power transmitted by the reader (P_{tran}) enabling the activation of the chip [BHA 10a, PER 11]*

As an example, if we take the RFID antenna shown in Figure 2.16, considering the two frequencies of 915 and 955 MHz, we can determine the minimal power required by the reader to activate the tag in various environments. We obtain these different environments by using the measurement protocol introduced in Chapter 2; that is by placing the tag on different materials with the same geometry but different permittivities. The results are shown in Figure 3.11. We can clearly see that the activation power is a function of the dielectric permittivity of the tag's support. Note that the sensor's sensitivity in the permittivity range from 2.2 to 5.7 is of interest; it is also a function of frequency. Likewise, this sensitivity could be improved if this criterion was integrated into the design of the tag's antenna. Note, however, that these results are a function of the measurement distance, which must be known and fixed throughout the entire measurement. A more general approach has been introduced by Marrocco [MAR 11]. With this method, the environment of the tag does not need to be *a priori* fixed or known. Indeed, it has been proved analytically and confirmed with measurements [CAI 11] that an analog ID independent on the position of the reader and the immediate environment can be obtained from the electromagnetic responses of the tag.

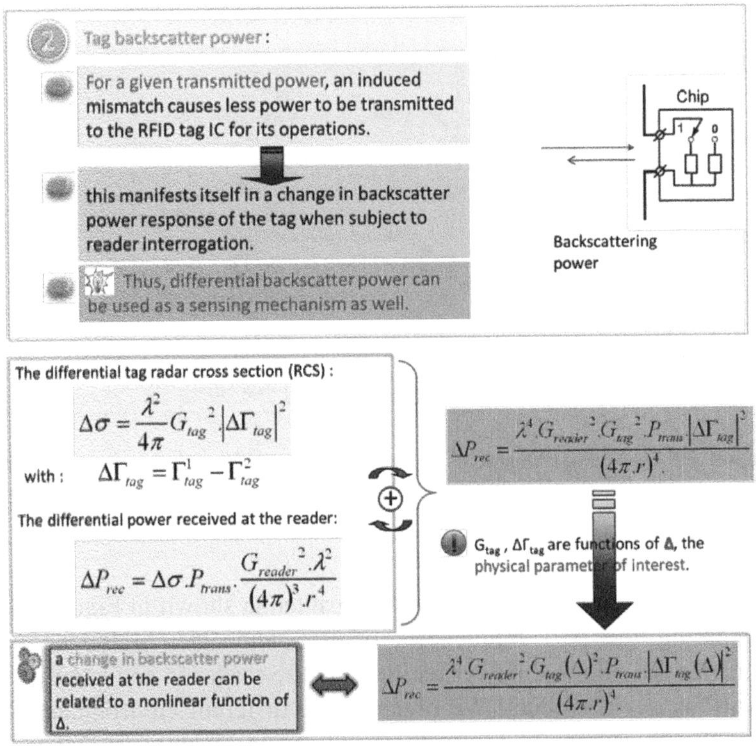

Figure 3.10. *Operating principle of a sensor based on RFID technology. In this second configuration, the information sought is given by the variation of power received by the reader (ΔP_{rec}) [BHA 10a, PER 11]*

Based on these different examples, we may wonder how to improve the sensitivity of the sensor function working solely on the antenna's design. This is motivated by the fact that we wish to remain as close as possible to the RFID system, and thus to continue to work with classic, currently available RFID chips. It is a complicated matter to answer this question directly, since sensor applications are so different (movement, temperature, liquid-level sensing, etc.). They will require specific research, most often directly connected to the performances of the chips used. However, we can introduce a fairly general approach when the information sought is related to the tag's RCS. In this case, it may be interesting, either in order to increase the sensitivity of the sensor function or to increase the read range for a given sensitivity, to attempt to optimize the RCS variation between the two impedance states of the chip. This is why we will look next at applying the RFID tag design technique described in Chapter 2, with the objective of increasing the ΔRCS of the tag.

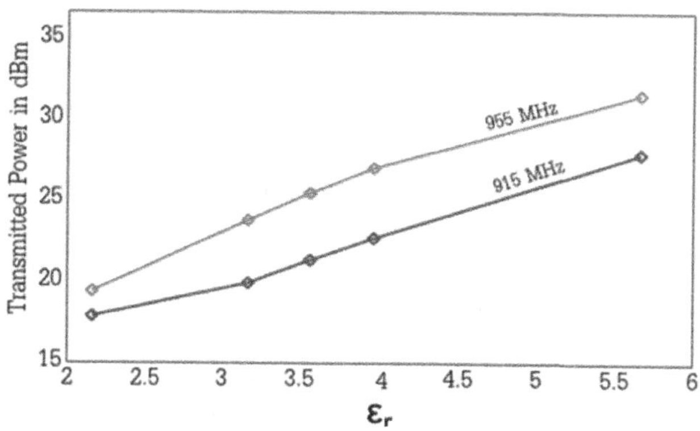

Figure 3.11. *Minimum tag activation power generated by the reader depending on the permittivity of the tag's support for two frequencies. The tag (see Figure 2.16) is placed on 1 mm-thick dielectric plates with permittivities covering the range 2.2–5.7. The measurement distance between the reader's antenna and tag is 1 m*

3.3.1. *Taking into account of downlink – increase of delta RCS*

The study presented here is based on the work of Professor Jean-Charles Bolomey, to whom we owe expression [2.4] [BOL 10]. As we have seen, the use of a tag as a sensor generally tends to reduce the read range (see Figure 3.11). Variations in the physical parameter being measured most often harm tag performance by modifying the matching condition between the antenna and chip. This is also the principle on which some of the approaches we have discussed are based (Figure 3.9). Based on this observation, it is preferable to attempt to use the variation in power received by the reader, also called the delta RCS (Figure 3.10), which is written interchangeably as either ΔRCS or Δσ in accordance with the notations previously introduced. We can see that in order to increase the measurement range, even for certain cases of sensitivity, it is of interest in terms of antenna design to maximize Δσ for a given RFID chip.

We will now look at how the design approach introduced in Chapter 2 can be used to respond to this issue. In this case, the two impedance states of the chip must be characterized with complete precision. In addition, the downlink must be taken into consideration, which is not the case for a classic tag design method. We saw in Chapter 2 that there is a compromise to be made between maximizing the power reaching the chip on the one hand, and maximizing Δσ on the other hand. The first specification consists of producing the conjugate matching condition ($Z_a = Z_c^{0\,*}$) as perfectly as possible, while the second specification consists of satisfying equation [2.4]. We will see in concrete terms to what this corresponds, and more exactly, how it is possible to find the best compromise between these two relationships [CHA 11].

We will use the NXP chip whose characteristics were given earlier (see Table 3.1) as our reference chip. Based on the chip values obtained in this way, we can show on a graph the point locations (in this case, the impedance values of the antenna Z_a, the real part, and the imaginary part) that will more or less satisfy each of the two expressions listed above. In order to better understand this representation, it is useful to reintroduce the power transmission coefficient for the absorbent state τ^0 and the modulation efficiency, written as [BOL 10]:

$$\tau^0 = 1 - \left|\Gamma_{tag}^0\right|^2,$$

$$ME = \left|\Gamma_{tag}^0 - \Gamma_{tag}^1\right|^2$$

with

$$\Gamma_{tag}^{i} = \frac{Z_c^i - Z_a^*}{Z_c^i + Z_a}, \; 0 \leq \tau^0 \leq 1 \text{ and } 0 \leq ME \leq 4$$

Using the chip impedance values given in Table 3.1, that is for a frequency of 915 MHz, if we look for the impedance value that maximizes the ΔRCS [2.4], we get the value below:

$$Z_{a,MEopt} = 61.15 + \text{j}126.5\,\Omega$$

For this antenna impedance value $Z_{a,MEopt}$, in this case we get an optimal modulation efficiency in relation to the chip considered: $ME_{opt} = -0.65\,dB$.

It is interesting to note that this value is much lower than the maximum value permitted ($ME_{max} = 4 = 6\,dB$). However, it is directly related to the intrinsic characteristics of the NXP chip (the values of the two impedance states Z_c^0, Z_c^1), and no action during the antenna design phase can be taken to improve it. It is, therefore, the best case scenario to reach this value. With regard to the matching condition, τ^0 may be equal to 1 (0 dB; that is the maximum value) for an impedance value: $Z_{a,\tau opt} = Z_c^{0\,*} = 25.6 + \text{j}147.7$. From there, Figure 3.12 shows the curves of the coefficients τ^0 and ME (expressed in dB) according to the tag antenna input impedance for the NXP chip. It is clear in the figure that the impedance domains Z_a offering the best modulation efficiency, as well as the best transmission coefficient, are close to one another. For example, within the shaded area shown in Figure 3.12, we simultaneously obtain $\tau^0 > -1$ dB and ME > -1dB. The optimal conditions: $Z_a = Z_{a,\tau opt}$ ($\tau^0 = 0$ dB, ME = 2.3 dB) and $Z_a = Z_{a,MEopt} = 61.15 + \text{j}126.52$ (τ^0 = -1 dB and $ME = ME_{opt}$= - 0.6 dB) are also shown in Figure 3.12 by points.

In order to obtain an antenna for which we are also able to take into account the modulation coefficient and thus the ΔRCS coefficient, the UHF RFID antenna design tool introduced in Chapter 2 is used. Besides taking into account the two impedance states in order to have access to the ΔRCS, we will also seek to obtain a tag that is robust to the environment, as was previously done. Two types of cost/fitness functions will also be used in order to best satisfy the following objectives:

– obtaining a tag with a large ΔRCS over the whole RFID UHF band;

– producing a tag with as low sensitivity as possible to dielectric disruptions.

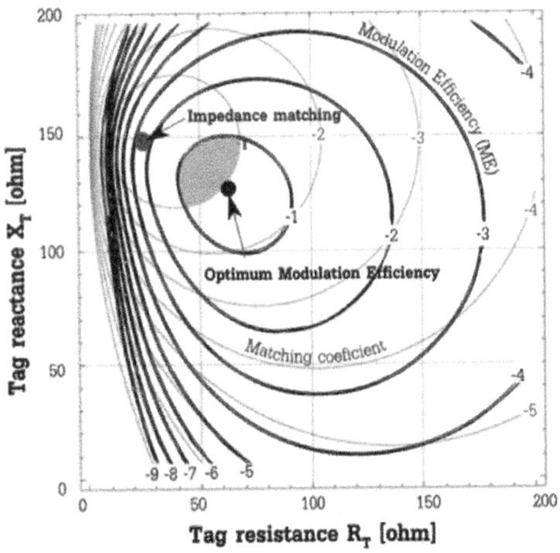

Figure 3.12. *Graphic representation of tag performance according to the input impedance value Z_a of its antenna (real part and imaginary part) for NXP chip at 915 MHz. The lines of the τ^0 (uplink performance) and ME (downlink performance) are traced for different values (in dB). The optimal conditions are represented by points*

In summary, the tag must be designed to have a good ΔRCS in free space, as well as when it is placed atop a dielectric material such as polytetrafluoroethylene (PTFE), with a relative permittivity of 2.2 and a thickness of 10 mm. In terms of frequency, these specifications are sought for the whole UHF band: 860–960 MHz.

The tag antenna best responding to these constraints after the use of the automated design tool is shown in Figure 3.13(a). The results obtained by measurement are shown in Table 3.2; they have been compared to the reference tag, specifically the tag NXP FFL 95-8, also operating with chip NXP G2XL. We can see the impact of the material on the read range, as well as on the ΔRCS. For relatively comparable read ranges, the tag designed makes it possible to have ΔRCS levels that are much higher than those measured with the NXP tag. Finally, we obtain a tag producing a good compromise between read range and ΔRCS. This behavior can clearly be observed in Figure 3.14.

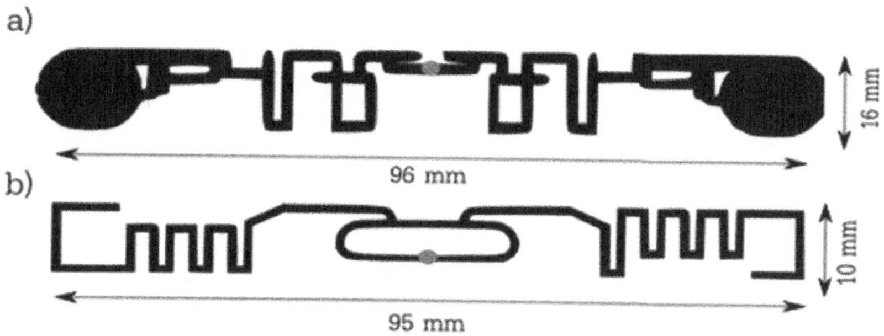

Figure 3.13. *Antennas for passive UHF RFID tags. a) Antenna produced that favors ΔRCS (also called modulation efficiency) in relation to the matching condition. b) Reference tag: NXP FFL 95-8*

Tag		EU		USA		JAPAN	
		Read range (m)	ΔRCS (cm^2)	Read range (m)	ΔRCS (cm^2)	Read range (m)	ΔRCS (cm^2)
Design tool	Air PTFE	2.8 7.2	46 290	4 6.3	92 107	5.5 4.5	249 54
NXPFF L 95-8	Air PTFE	6.7 5.8	26 32	5.9 6.6	60 11	6.15 6.8	23 5

Table 3.2. *Read ranges and ΔRCS measured for the tag designed to have a large ΔRCS (Figure 3.13(a)), as well as for the NXP tag, over the three RFID frequency bands*

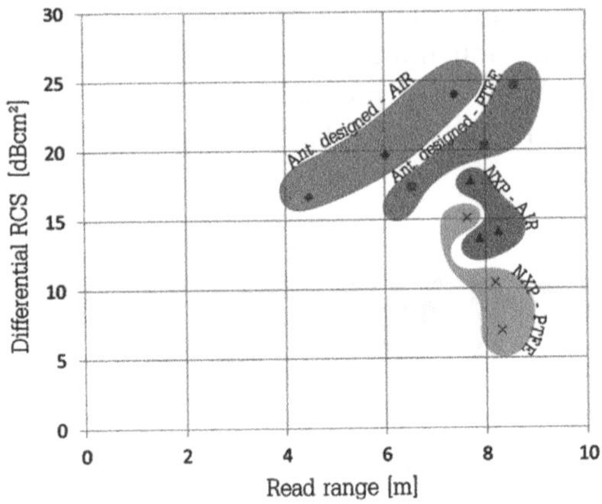

Figure 3.14. *Representation of ΔRCS measurement according to read range for optimized tag and reference tag (NXP)*

We can see that the information obtained using the RFID chip characterization method (presented at the beginning of this chapter) can be used to design an antenna adapted to the two states of the chip over the whole RFID UHF frequency band. The variation of the RCS can thus also be taken into consideration. The increase of this variation range will also enable an increase in performance of the sensor function associated with this tag.

To conclude this chapter, we will give an example of an RFID sensor in order to illustrate our suggestions concretely. Emphasis will be placed above all on the simplicity of using a classic RFID tag as a sensor. This example will also allow us to note the limitations that can be observed in equivalent scenarios.

3.3.2. *Example of an RFID sensor*

We will now present the approach that can be applied in order to use a classic RFID tag to detect the level of a liquid in a plastic container. The specifics of the measurement taken are described in Figure 3.15. A standard UHF tag has been reused as a wireless sensor. For the reader part, an arbitrary waveform generator (transmission of the RFID frame) and a spectrum analyzer (detection of the tag's ID) are used to detect the activation power of the RFID tag (Figure 3.15(a)). This is done for different water levels in a plastic container (Figure 3.15(b)). The results shown in Figure 3.16 show that the activation power is a function of the water level. The higher the water level, the more the antenna of the tag is mismatched, and thus a higher level of power must be sent to it. We can also see that the relationship between the water level and power is not linear. Moreover, the maximum water level that can be detected is highly dependent on the distance between the tag and the reader's antenna. In fact, for a read range of 50 cm, it is not possible with the generator (17 dBm maximum at output) to measure a water level higher than 2 cm. We can see here the first limitations of the approach; specifically, that it is highly sensitive to the tag's environment, as in this case to the read range, for example. These parameters will be involved both in terms of sensitivity and total measurement range, and these limitations will have all the more impact when we seek to measure very slight variations. In this case, imprecisions and measurement errors related to the RFID chip will become too influential.

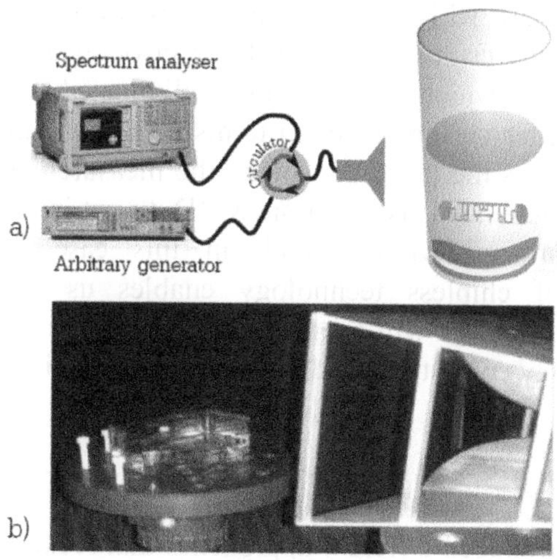

Figure 3.15. *Measurement bench used to detect the level of liquid in a plastic container using a classic RFID system (tag + reader). The reader is emulated here using an arbitrary waveform generator as well as with a spectrum analyzer. a) Theoretical schema of bench. b) Photo of bench in anoechoic chamber*

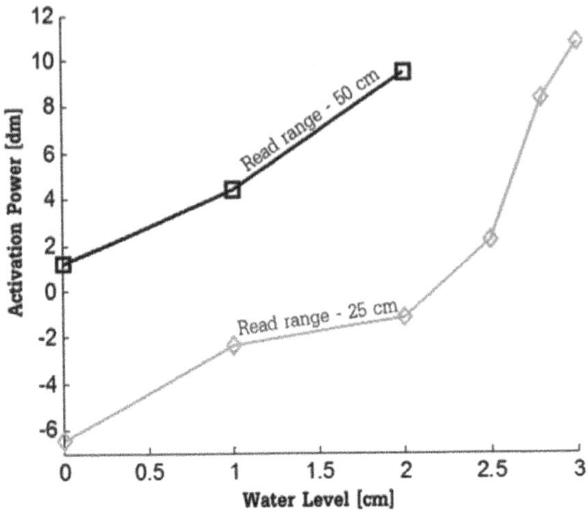

Figure 3.16. *Measurements of tag activation power (here the power sent by the generator) depending on the water level present in the plastic container. Measurements were taken for two distances between the antenna and the water-filled container*

In general, UHF RFID chips behave in ways that are difficult to reproduce, which results in the need for error measurement. From there, a solution in which the chip can be eliminated will enable us to obtain much better performances in terms of sensor function. This is not possible unless the tag used to take the measurement also has its own ID. We would need, as in classic RFID, to put in place a network of sensors that are wireless and, in this case, chipless. The development of chipless technology enables us to resolve this difficulty perfectly. We will see in the next chapters what this fairly recent technology consists of, and its potential in applicative terms.

3.4. Conclusion

We have introduced different approaches that seek to associate the identification methods already present in RFID with functions to measure or detect physical values. The principal advantage of these approaches lies in the fact that they rely fully on the use of classic RFID systems (passive UHF in this case), to which we will add the desired functionality without greatly modifying the tag or the reader. Thus, we will rely solely on the RFID communication principle, which remains a widely implemented standard and thus one that is easy to use. We have seen through several examples the potential of this approach, particularly the possibility of using it to measure the input impedances of small antennas. Based on this technique, wireless and batteryless sensors, which are potentially low-cost and have a long lifespan, have been produced. However, it should also be noted that only the concept proof has been accomplished to date. We are still far from being in a position to implement industrial applications, even though these "free" functionalities (or in any case requiring only slight technological developments) may appear highly attractive from an applicative point of view. In addition, it must be admitted that classic RFID chips cannot really be considered precision components, which will unavoidably limit their domains of use as sensors.

Next, we will examine a new concept for RFID tags, this time chipless. An approach relatively comparable to the one described here may also be used. In this case, the absence of the chip, and thus of some imprecisions related to it, will be eliminated, making it possible to obtain tag-sensors that can be very precise in measuring some parameters, such as humidity level or surface distortion.

3.5. Bibliography

[AND 14] ANDIA VERA G., DUROC Y., TEDJINI S., "RFID test platform: nonlinear characterization", *IEEE Transactions on Instrumentation and Measurement*, vol. 63, no. 9, pp. 2299–2305, 2014.

[BHA 09] BHATTACHARYYA R., FLOERKEMEIER C., SARMA S., "Towards tag antenna based sensing – an RFID displacement sensor", *2009 IEEE International Conference on RFID*, Orlando, Florida, US, pp. 95–102, 2009.

[BHA 10a] BHATTACHARYYA R., FLOERKEMEIER C., SARMA S., "Low-cost, ubiquitous RFID-tag-antenna-based sensing", *Proceedings of the IEEE*, vol. 98, no. 9, pp. 1593–1600, 2010.

[BHA 10b] BHATTACHARYYA R., FLOERKEMEIER C., SARMA S., "RFID tag antenna based sensing: does your beverage glass need a refill?", *2010 IEEE International Conference on RFID*, Orlando, Florida, US, pp. 70–77, 2010.

[BHA 11] BHATTACHARYYA R., FLOERKEMEIER C., SARMA S., *et al.*, "RFID tag antenna based temperature sensing in the frequency domain", *2011 IEEE International Conference on RFID (RFID)*, Orlando, Florida, US, pp. 70–77, 2011.

[BOL 10] BOLOMEY J.C., CAPDEVILA S., JOFRE L., *et al.*, "Electromagnetic modeling of RFID-modulated scattering mechanism. Application to tag performance evaluation", *Proceedings of the IEEE*, vol. 98, no. 9, pp. 1555–1569, 2010.

[BOR 10] BORIES S., HACHEMI M., KHLIFA K.H., *et al.*, "Small antennas impedance and gain characterization using backscattering measurements", *2010 Proceedings of the 4th European Conference on Antennas and Propagation (EuCAP)*, Barcelona, Spain, pp. 1–5, 12–16 April 2010.

[CAI 11] CAIZZONE S., MARROCCO G., "RFID grids: Part II – experimentations", *IEEE Transactions on Antennas and Propagation*, vol. 59, no. 8, pp. 2896–2904, 2011.

[CHA 11] CHAABANE H., PERRET E., DAIKI M., *et al.*, "RFID tag design approach based on chip multi states impedance", *Proc. Asia-Pacific Microwave Conference (APMC)*, Melbourne, 5–8 December pp. 1466–1469, 2011.

[CHE 09] CHEN S.-L., LIN K.-H., "Characterization of RFID strap using single-ended probe", *IEEE Transactions on Instrumentation and Measurement*, vol. 58, no. 10, pp. 3619–3626, 2009.

[DAI 11] DAIKI M., CHAABANE H., PERRET E., *et al.*, "RFID chip impedance measurement for UHF tag design", *Progress in Electromagnetics Research Symposium (PIERS'11)*, Marrakesh, Morocco, pp. 679–684, 2011.

[GRE 63] GREEN R.B., The general theory of antenna scattering, Doctor of Philosophy, Ohio State University, Electrical and Computer Engineering, 1963.

[GUI 12] GUILLET A., VENA A., PERRET E., *et al.*, "Design of a chipless RFID sensor for water level detection", *15th International Symposium on Antenna Technology and Applied Electromagnetics (ANTEM)*, Toulouse, France, pp. 1–4, 2012.

[HAI 10] HAILONG Z., KO Y.C.A., YE T.T., "Impedance measurement for balanced UHF RFID tag antennas", *IEEE Radio and Wireless Symposium (RWS)*, pp. 128–131, 2010.

[HAN 89] HANSEN R., "Relationships between antennas as scatterers and as radiators", *Proceedings of the IEEE*, vol. 77, no. 5, pp. 659–662, 1989.

[HAR 63] HARRINGTON R., "Field measurements using active scatterers", *IEEE Transactions on Microwave Theory and Techniques*, vol. 11, no. 5, pp. 454–455, 1963.

[HAR 64] HARRINGTON R.F., "Theory of loaded scatterers", *Proceedings of the Institution of Electrical Engineers*, vol. 111, no. 4, pp. 617–623, 1964.

[HAR 96] HARRINGTON R.F., HARRINGTON J.L., *Field Computation by Moment Methods*, Oxford University Press, 1996.

[LEO 07] LEONG K.S., NG M.L., COLE P.H., "Investigation of RF cable effect on RFID tag antenna impedance measurement", *Antennas and Propagation Society International Symposium, IEEE*, pp. 573–576, 2007.

[MAN 12] MANZARI S., OCCHIUZZI C., NAWALE S., *et al.*, "Polymer-doped UHF RFID tag for wireless-sensing of humidity", *2012 IEEE International Conference on RFID*, pp. 124–129, 2012.

[MAR 08] MARROCCO G., MATTIONI L., CALABRESE C., "Multiport sensor RFIDs for wireless passive sensing of objects – basic theory and early results", *IEEE Transactions on Antennas and Propagation*, vol. 56, no. 8, pp. 2691–2702, 2008.

[MAR 11] MARROCCO G., "RFID grids: Part I – electromagnetic theory", *IEEE Transactions on Antennas and Propagation*, vol. 59, no. 3, pp. 1019–1026, 2011.

[MAY 08] MAYER L.W., SCHOLTZ A.L., "Sensitivity and impedance measurements on UHF RFID transponder chips", *2nd International EURASIP RFID Technology Workshop*, Budapest, Hungary, pp. 1–10, 2008.

[MAY 94] MAYHAN J.T., DION A.R., SIMMONS A.J., "A technique for measuring antenna drive port impedance using backscatter data", *IEEE Transactions on Antennas and Propagation*, vol. 42, no. 4, pp. 526–533, 1994.

[MEY 98] MEYS R., JANSSENS F., "Measuring the impedance of balanced antennas by an S-parameter method", *IEEE Antennas and Propagation Magazine*, vol. 40, no. 6, pp. 62–65, 1998.

[NIK 06] NIKITIN P.V., RAO K.V.S., "Theory and measurement of backscattering from RFID tags", *Antennas and Propagation Magazine, IEEE*, vol. 48, no. 6, pp. 212–218, 2006.

[NXP 08] NXP Application Note, AN 1629 UHF RFID Label Antenna Design, UHF Antenna Design, September 2008.

[NXP 12] NXP, Product data sheet, SL3ICS3001 UCODE HSL, 2012. Available at http://www.nxp.com/documents/data_sheet/SL3ICS3001_072831.pdf.

[PER 11] PERRET E., TEDJINI S., "RFID tags: from identification to sensing", *11th Mediterranean Microwave Symposium MMS'2011*, Hammamet, Tunisia, 2011.

[PUR 08] PURSULA P., SANDSTROM D., JAAKKOLA K., "Backscattering-based measurement of reactive antenna input impedance", *IEEE Transactions on Antennas and Propagation*, vol. 56, no. 2, pp. 469–474, 2008.

[QIN 09] QING X., GOH C.K., CHEN Z.N., "Impedance characterization of RFID tag antennas and application in tag co-design", *IEEE Transactions on Microwave Theory and Techniques*, vol. 57, no. 5, pp. 1268–1274, 2009.

[SID 07] SIDEN J., ZENG X., UNANDER T., *et al.*, "Remote moisture sensing utilizing ordinary RFID tags", *IEEE Sensors*, pp. 308–311, 2007.

[WIE 98] WIESBECK W., HEIDRICH E., "Wide-band multiport antenna characterization by polarimetric RCS measurements", *IEEE Transactions on Antennas and Propagation*, vol. 46, no. 3, pp. 341–350, 1998.

PART 2

Chipless RFID

4

Introduction to Chipless RFID

4.1. Introduction

Data exchange systems using radio frequency identification (RFID) technology are commonly used to recognize and/or identify, over small or medium distances, any type of object bearing a specific tag. It is possible, using this technique of automatic information capture via radio reading, to collect information remotely and without contact. There are various types of RFID tags; we are distinguishing here between tags including an integrated electronic circuit, called RFID tags (discussed in Part 1 of this book); and tags that do not include an integrated electronic circuit, generally called chipless tags or Radio frequency (RF) barcodes. This latter family of tags will be the subject of this chapter; more precisely, we will introduce this technology, which appeared in the mid-2000s.

To understand the development of chipless RFID, it is important to view it in comparison to classic RFID and barcode. We can see that RFID benefits from a very wide functional spectrum, related to the use of RF waves for data exchange. Here, the acquisition of the identifier is made much easier and volumetric readings are possible, all on tags containing modifiable information. These functions are impossible to implement with a barcode, and we might think that the latter would be inevitably overtaken; and yet, according to the figures, nothing of the kind has happened. In reality, 70% of the items manufactured worldwide are equipped with it. The reasons for this enthusiasm are simple: it functions very well and both the label and the reader are extremely cheap. This is why barcodes remain the uncontested benchmark in terms of identification, with a cost-to-simplicity-of-use ratio that remains unequalled. It is also true that classic RFID contributes to other significant functionalities, and the question is therefore one of imagining a technology based on RF waves as a communication vector that would retain some of the advantages of barcodes. Pragmatically speaking, the question of system cost, and particularly of the tags that must be produced in large numbers, remains the central point. Due to the presence of electronic circuits, these tags have a non-negligible cost much higher than that of barcodes. Therefore, it is logical that a simple solution consists of producing chipless RFID tags. The high cost of RFID tags is actually one of the principal reasons that chipped RFID is rare in the market for widely distributed products, a market that numbers in the tens of thousands of billions of units sold per year. In this market, optic barcodes are very widely used. However, technically speaking, chipped RFID offers significant advantages including increased reading distance and the ability to detect a target outside the field of vision, whatever its position. As we will see, the concept of chipless RFID is a solution to this issue, and should eventually compete with barcodes in certain areas of application.

Generally speaking, chipless tags are most often composed of a standard substrate on which conductive patterns are placed, the specific geometry of which makes it possible to identify the tag with certainty. Depending on the approach used, these devices may or may not possess a ground plane. The information coding principle; that is, the tag identification code is based on the generation of a specific electromagnetic (EM) signature, somewhat like the principle of radar. The main difference here is that the form of the conductive pattern is imposed in order to have a specific and perfectly recognizable signature. Thus, the information is no longer memorized using an electronic chip as in traditional RFID tags, but rather directly "inscribed" in the tag. Consequently, an essential point lies in the link that exists between the geometry of the conductive pattern and the RF signature expected. At the applicative level, it must actually be possible to determine the tag's pattern directly using the identifier being coded. Considered from this angle, this point goes back in a way to the attempt to resolve an inverse problem that may be complicated.

Chipless tags most often offer mechanical and thermal reliabilities superior to chipped tags. Likewise, they require a "feed" power (the power transmitted by the reader) lower than that of standard RFID tags. Unlike barcode or magnetic strip applications, it might be unnecessary to orient (or position) the tag in a specific direction or place. Reading takes place remotely, without direct visibility; that is, through most opaque, non-metallic materials. However, like barcodes, this technology does not currently offer the possibility of rewriting the data in the tag (read only). In a way, the information contained in the tag can be considered as non-modifiable. After having described the operating mode of chipless RFID in detail, we will see that in terms of targeted applications, chipless technology falls midway between classic RFID and the use of barcodes.

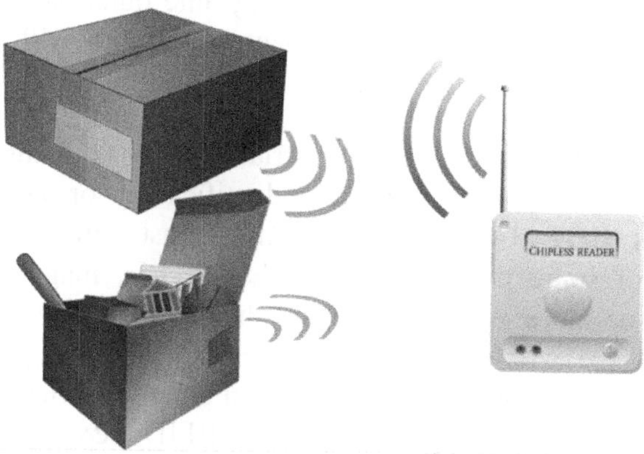

Figure 4.1. *Operating principle of a chipless RFID system. The tag is composed of a conductive pattern, possibly printed on a paper or polyethylene terephthalate (PET) substrate*

4.2. Operating principle of chipless RFID

Like various existing RFID technologies, chipless RFID tags are associated with a specific RF reader, which questions the tag and recovers the information contained in it. The operating principle of the reader is based on the emission of a specific EM signal toward the tag, and the capture of the signal reflected by the tag (see Figure 4.1). The processing of the signal received – notably via a decoding stage – makes it possible to recover the information contained in the tag. However, chipless tags are fundamentally different from RFID tags. In the latter, a specific frame is sent by the reader toward the tag according to a classic binary modulation schema. The tag demodulates this signal, processes the request, possibly writes data in its memory, and sends back a response, modulating its load. Chipless RFID tags, however, function without a communication protocol. They can be viewed as radar targets possessing a specific, stationary temporal or frequential signature. With this technology, the remote reading of an identifier consists of analyzing the radar signature of the tag. This stationary character is very important in practice, since in order to improve the signal-to-noise ratio, or to avoid outside disturbances from other EM signals, here it is sufficient to proceed with an averaging of the backscattered signal, repeating the tag's interrogation sequence. Because the tag backscattering time duration is extremely short (on the order of a few dozen nanoseconds), it will not have a substantial effect on reading time.

Much academic and even industrial work has focused on the development of chipless tags for identification or authentication applications [IDT 14, MET 02]. They most often use specific properties of some particular materials (electric, magnetic, or EM properties, or more widely physical, chemical or physico-chemical properties). We will limit ourselves here to describing approaches based solely on the use of RF waves which greatly restricts the field. It is for this reason that we speak of chipless RFID; here, the term RFID indicates that we are seeking to identify a chipless tag with an EM signal.

The earliest works on chipless RFID include the work of Leon Theremin [WIK 14a], who in the 1940s created a totally passive chipless system for remote listening [NIK 12]. This passive listening device became known as the Great Seal bug [WIK 14b]. It is in some sense an ancestor of RFID, both chipless RFID and today's sensor tags. We may also cite the development of a detection system known as Electronic Article Surveillance (EAS) [WIK 14c]. EAS tags are also chipless and provide a simple means of detecting a tag when it is near a reader. All these tags have the same information and operate with magnetic fields at low-frequency (in the 10 – 1,000 Hz range). The operating principle is based on the use of nonlinear magnetic material which enables the generation of higher harmonics that are easier for the reader to detect.

In the early 2000s, work conducted at the Massachusetts Institute of Technology (MIT) also focused on the theme of chipless RFID. In [FLE 02] we can see an overview of near-field EM tagging. Chipless tags operating in the high frequency (HF) frequency range, and based on multiply-resonant planar metal structures have been implemented. This thesis entitled "Low-Cost Electromagnetic Tagging: Design and Implementation" is particularly interesting in that it introduces a certain chipless vision that is still followed today. It also describes a method of making relatively low-cost tags. Unlike these approaches, the studies presented below show the advantages of using RF waves and tags for reader communications located in the far field. This gives some reading flexibility, particularly by enlarging the reading zone. This new technological approach has stimulated new research in this area, which will be introduced later.

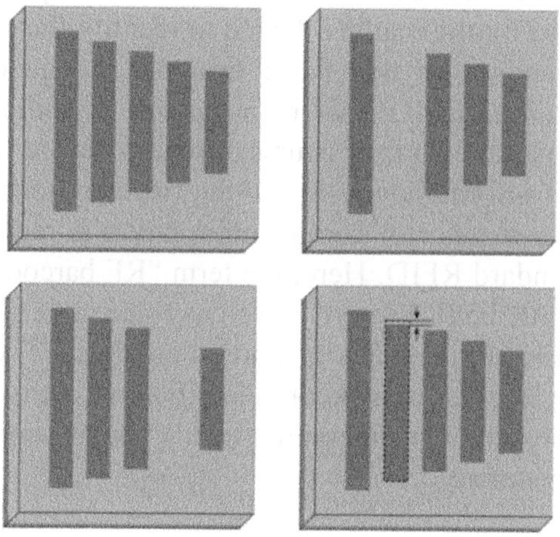

Figure 4.2. *Example of a chipless dipole tag with a ground plane [JAL 05a]. The tag's dimensions are 18 × 35 mm and around 1 mm thick*

4.2.1. *Description of the principle of chipless RFID*

Figure 4.2 represents a basic example of a chipless RFID tag. This structure, which use multiple half-waves, short-circuits, and dipoles was one of the first chipless RFID tags introduced and able to operate in the far-field zone [JAL 05a]. The tag is formed from a dielectric substrate whose rear surface is covered by a metallic ground plane. On the upper face of the substrate are five parallel conductive bands defining 5 dipoles of varying length.

The tag thus constitutes a structure with scatterers liable to interfere with an EM wave generated by a reading device. Each dipole acts as a resonant circuit, re-emitting a part specific to the incident signal. The reflected wave can be sensed by the reading device. Thus, each conductive strip determines, via its geometry, a resonance frequency of the tag. In operation, the reading device generates a radio signal whose spectrum includes all of the resonance frequencies of the tag it is liable to read. If the tag is near the reading device, this device detects an *extremum* of the signal at the resonance frequencies determined by the lengths of the dipoles, which is manifested by the appearance of five distinct elements in the power spectrum of the radio signal. The positions of these five distinctive elements in frequency (peaks or troughs) are used by the reading device to identify the tag in a unique manner. We can see in this basic example the general operating principle of chipless RFID; that is, a method of coding information using a totally passive and, of course, chipless device. Note also that the exchange of information relies on the two-way journey of an EM wave, where the manner of coding information is radically different from that of standard RFID. Here, the term "RF barcode" takes on its full meaning [JAL 05a]. Likewise, we can make a very clear analogy with the principle of the radar signature of an object. The difference in this case is that in order to make an identification we must be capable of linking the object's EM signature directly and as simply as possible to its intrinsic geometry.

Let us go back now to the work of design, which consists of developing tags with a one-to-one link between their geometries and the backscattered EM signature. It is clear that in terms of use, it is not possible to simulate all the different configurations of a tag at the same time (after the point at which the tag presents more than 10 bits of information), in order to be able to associate them subsequently with an identifier. The user must be able to obtain the tag's geometry directly from the identification number he/she wishes to affix to the tag. To do this, it is possible to use relatively simple considerations, such as those based on resonant circuits. In fact, if we take the example of the short-circuited dipole previously introduced, by comparing the signal received to the signal sent we can detect its resonance frequency. This example clearly shows that there is a direct link between geometry (length in this case) and resonance frequency (and therefore, in the end, the frequency of the tag's code). By changing the length of the dipole we can shift the position of the separate elements inside a defined frequency band. However, there are several limitations to this. To increase the tag's coding capacity, that is, the number of different configurations (various geometries and thus resonance frequencies), it appears to be of interest to increase the number of resonators, that is, the number of dipoles, as shown in Figure 4.2. However, to maintain a direct link between code and geometry (in this case the length of each dipole), the resonators must be uncoupled from one another. In the opposite case, the geometric modification of a length, besides impacting the resonance frequency associated with it, would also affect the frequencies of the neighboring dipoles, rendering the device unusable for making an identification. A complete uncoupling is, unfortunately, not possible, all the more so because there are strict limits on tags' dimensions (the idea being to minimize their surface area). For this reason, this coupling phenomenon makes it necessary to associate a frequency interval with each family of chipless tags, from which it is possible to determine whether the frequential discrepancy observed is truly the result of a change of configuration, and thus of code.

The second limitation arises from the frequency band. It is also important, due to the restrictive character that exists in terms of the usable frequency band, to work with resonators that have a significant quality factor. Figure 4.3 shows various types of resonators that could potentially be used to produce chipless tags. These include the dipole in short-circuit (Figure 4.3(c)), in slot form (Figure 4.3(a)), and in a slot line that is short-circuited on one side and open on the other

(Figure 4.3(b)). Figure 4.4 shows the radar equivalent surfaces associated with the three types of resonators in Figure 4.3. Above all, we can see that the presence of distinctive elements (peaks or troughs) at predefined frequencies can be used to code a specific identifier. Moreover, we can also see that the half-wave dipole has a low quality factor, and its radar cross-section (RCS) is of the order of -24 dBsm. The first slot resonance also occurs every half-wavelength, but its selectivity is significantly improved, as is its RCS (-18.5 dBsm). In contrast, the C-shaped structure is a quarter-wave resonator with a significant quality factor. This latter basic structure offers a certain advantage of miniaturization, and thus of coding density. It clearly seems that the choice of resonator will determine the tag's capacity, as well as its final dimension [VEN 13]. For slot- and C-type structures (more generally for no-ground simple scatterers, as explained in [VEN 13]), we observe the presence of a peak followed by a trough on the RCS curve. Either one of these perfectly recognizable elements can be used to code information. In the following example, for the sake of ease, we will consider the trough's frequency position.

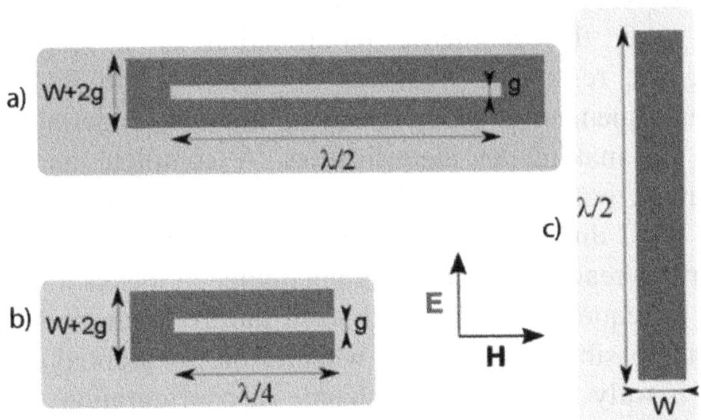

Figure 4.3. *Various types of resonators for the production of chipless tags: a) slot, b) C-shaped structure, c) short-circuited dipole*

With regard to the work of the designer, another approach is also possible. Rather than designing scatterers whose resonance frequencies will be adjusted within a given range based on a specific geometry, it is also possible to adopt an all-or-nothing approach (in term of classical modulation, it is similar to the well-known on–off keying (OOK) approach that represents digital data as the presence or absence of a carrier wave). In this case, we design a set of N resonators having, for example, resonance frequencies that are regularly spaced and cover the entire frequency band allowed. Each resonator is optimized to have a specific resonance frequency, bringing as many geometric parameters as required into play. From there, in order to code information, we simply allow or prevent each scatterer from resonating individually. Thus, to eliminate one or more N scatterers, it is a simple matter of not etched the resonator in question or, depending on the type of resonator used, short-circuited that resonator so as to eliminate its resonance frequency [PRE 08b, VEN 12c].

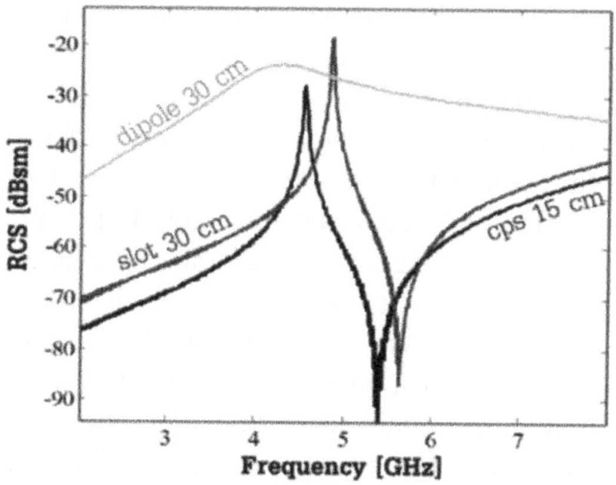

Figure 4.4. *RCS created by a short-circuited dipole (as in Figure 4.3), a slot (as in Figure 4.3) and a C-shaped structure (as in Figure 4.3(b)). Dimensions: $\lambda/2$=30 cm, W=2 mm, g=0.5 mm, with no ground plane*

The approach described here is only one of several. We will focus in this introductory section on the so-called frequential approach, which is related to the example of dipoles. As we will see, other approaches do exist; notably those in which the temporal discrepancy between multiple parts of signals is brought into play. Examples of these other methods will be given in later chapters. However, for further explanation we refer the reader to [TED 12], where a general study on the different ways that exist of coding information in chipless technology is discussed.

Now that we have introduced the basic principle of chipless RFID through a simple example, we will focus on an optimized device that definitively demonstrates the task of design as well as the compromises to be made with regard to the tag's coding capacity and dimension as well as the frequency band to be used. Likewise, specific attention will be paid to the comparison of coding densities that can be obtained depending on the approach used.

4.2.2. *Example of C-shaped tag*

Tags are composed of an ordinary substrate with no ground plane, on which conductive C-shaped strips are placed in an overlapping configuration, as shown in Figure 4.5. This specific topology is at the root of a large number of resonance frequencies, which are the basis of information coding. The resonances are directly dependent on the geometry of the conductive patterns. This particular arrangement forms slots of different lengths that resonate at specific frequencies. The modification of slot lengths (for example, by changing the length of short-circuits positioned at the slots) makes it possible to control the resonance frequencies very precisely, and thus to control the form of the EM signal sent back to the reader by the tag. As with the principle of radar reconnaissance, the tag thus described acts simultaneously as a receptor, a filter and a transmitter. The tag can be seen as a very specific radar target; that is, a target in which it is possible to precisely set the parameters of certain parts of its EM signature using its own geometry.

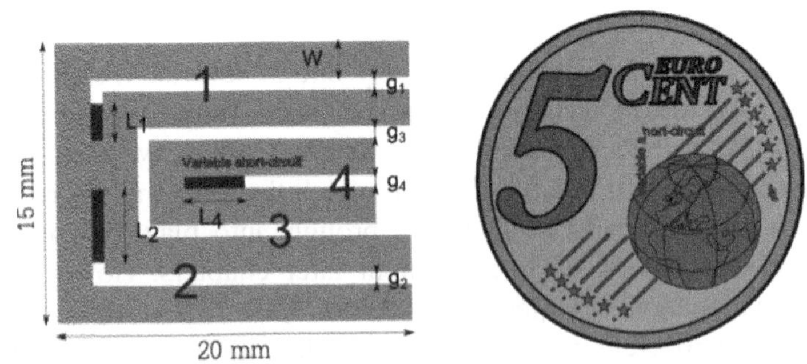

Figure 4.5. *C-shaped tag*

Consider the tag shown in Figure 4.5. It is composed of 4 resonators labeled 1, 2, 3 and 4, and independent of one another except for number 3, which allows the decoupling of resonators 1 and 2 on the one hand and from resonator 4 on the other hand [VEN 12b]. Short-circuits can be used to adjust the length of the three slots 1, 2 and 4, and thus the resonance frequencies. As previously indicated, the interest of this type of structure lies in the fact that these three resonances are totally uncoupled from one another, even though the slots are very close to each other. This absence of coupling makes it possible to control each resonance frequency independently, and therefore code a large amount of information on a surface reduced to 1.2 cm × 2 cm. This principle is illustrated in Figures 4.6 and 4.7. Figure 4.6 shows the three resonances relative to each of the three slots used, for three characteristic frequencies; only the length L_2 of one of the three slots varies (slot no. 2). We can see a difference in the resonance frequency corresponding to this slot, while the two others remain unchanged; this shows the very good uncoupling achieved between the slots.

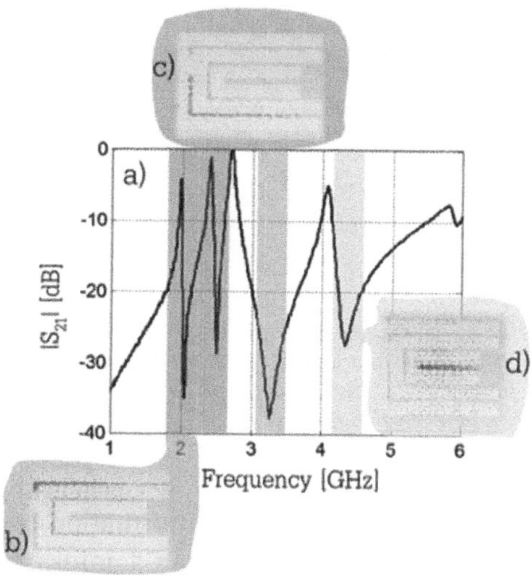

Figure 4.6. *Illustration of the link that exists between the resonance frequencies and the geometry of C-shaped tags. a) Amplitude of the signal backscattered by the tag according to frequency; the signal is normalized in comparison to the incident signal, b) current density on the tag at a frequency of 2.1 GHz, c) 2.55 GHz and d) 4.4 GHz. For a color version of this figure, see www.iste.co.uk/perret/radio.zip*

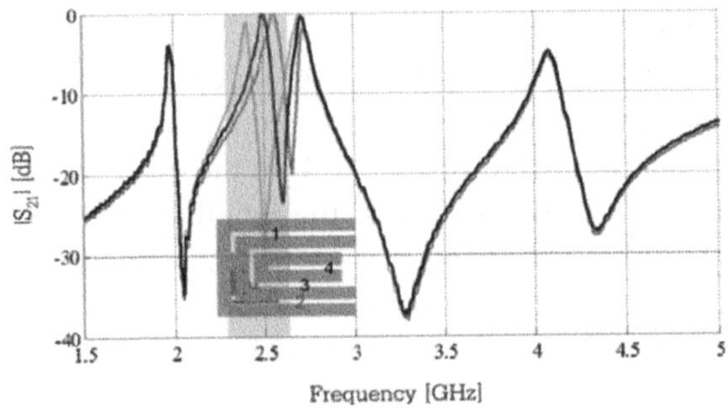

Figure 4.7. *Illustration of the operating principle of a C-shaped tag. Amplitude of the signal backscattered by the tag according to frequency for three different lengths L_2 of slot no. 2; the signal is normalized in comparison to the incident signal*

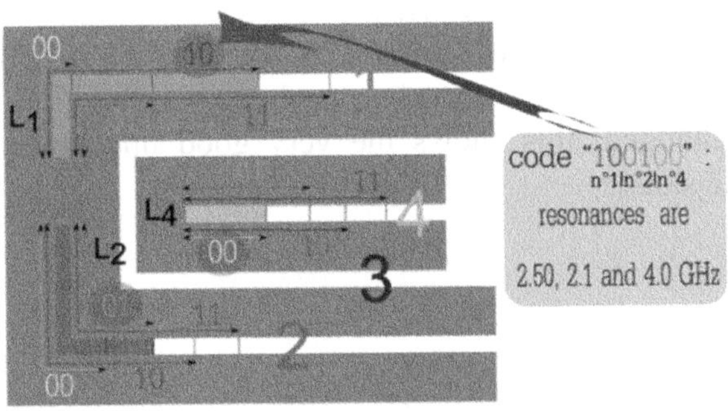

Freq. [GHz]	Code for slot N°1
2.40	00
2.45	01
2.50	10
2.55	11

Freq. [GHz]	Code for slot N°2
2.0	00
2.1	01
2.2	10
2.3	11

Freq. [GHz]	Code for slot N°3
4.0	00
4.4	01
4.8	10
5.2	11

Figure 4.8. *Illustration of the coding principle via the example of a C-shaped tag. The table of correspondences shows the link between the resonance frequencies of the tag and the associated binary code; here a code on 6 bits. For a color version of this figure, see www.iste.co.uk/perret/radio.zip*

Now let us focus on the recovery of information contained in a tag from its EM signature. To access the tag's information, the signal backscattered and recovered by the reader is processed. In this example, the frequencies for which the amplitude of the signal is minimal are extracted. Next, a correspondence between these physical values and a binary code (the tag ID) is defined, as shown in Figure 4.8. This correspondence may, for example, take the form of a table with two entries. This type of table is used to associate each combination of frequencies with a combination of zeros and ones, with the latter constituting the binary code of the tag. Figure 4.8 illustrates the coding principle implemented to link the signature to the binary code. In the example shown in this figure, we assume that each of the three slots used can have four different lengths, by modifying the lengths of the short circuits. Each slot length corresponds to a specific and known resonance frequency (the values of these frequencies are given in the table of correspondences), and we associate a two-bit binary code with each frequency. Insofar as the tag accommodates three totally uncoupled slots, we can code a total of 6 bits in this way, two bits per slot. As an example, if we are seeking to code the identifier 100100, according to the correspondence table we must make slot 1 resonate at 2.5 GHz, and slots 2 and 4 resonate at 2.1 and 4.0 GHz, respectively.

As stated above, there is a clear link between resonance frequencies and slot lengths. As we can see in Figure 4.9, it is therefore possible to model and trace the length of slots according to their resonance frequency. However, this work often requires specific study.

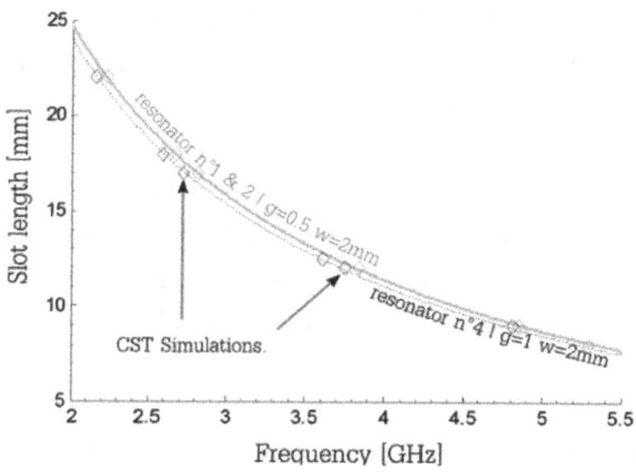

Figure 4.9. *Variation of slot lengths Ls according to their resonance frequency. The permittivity is 4.6 and the height of the substrate 0.8 mm. (__) Evolution of Ls for resonators 1 and 2 with g = 0.5 mm, w = 2 mm. (---) Evolution of Ls for resonator 4 with g = 1 mm and w = 2 mm. (◊◊◊) CST simulations obtained for specific lengths*

This chipless RFID tag is extremely compact, with a coding capacity of up to 10 bits (assuming the sensitivity of the reader [VEN 12b]) for a reduced surface area of 1.5 × 2 cm². Note that the compactness of the tag remains an important point, as chipless tags suffer from low storage capacity compared to standard RFID tags or even barcodes. The ratio of the number of bits per square centimeter of the tag described here is 3, which is much larger than for the one presented in Figure 4.2 [JAL 05a], which has a ratio of only 0.38 bits/cm².

Now that we have described the operating principle of a chipless tag, let us return to the positioning of this approach compared with other technologies of classic approaches (UHF RFID, barcodes) and even other chipless approaches, some of which are not necessarily of the RF type.

4.3. Positioning of chipless RFID

4.3.1. *Latest developments*

We will see here how the development of chipless RFID tags with RF frequencies has flourished significantly in recent years [PRE 10]. The appearance of chipless tags is intended to reduce the cost of RFID tags, and in addition to reducing chip prices, this technique also eliminates costs related to the positioning and connection of these chips in relation to the antenna; these elements represented up to two-thirds of the price of the tag [PER 13].

Chipless tags, whether RF or not, are usually devices formed of low-cost components, magnetic materials, or, very generally, materials that reflect or absorb RF waves. Chipless devices, compared to passive chipped RFID tags, are normally characterized by the following:

– lower price;

– low size factor. The devices are made from thin materials or films;

– low-capacity data;

– more reliable functioning and thermal and mechanical control much greater than that of chipped tags;

– very low requirements in terms of power transmission from the reader.

This last point is an important one; in chipless RFID, the power transmitted by the reader can be much lower than the power used in the case of standard RFID tags. In fact, since these tags do not contain a chip, it is not necessary to provide a certain level of power to make them operate. In the case of chipless technology, it is the sensitivity of the reader that generally determines the operating distance.

Note, however, that of the characteristics listed above (some of which are clearly advantageous), the low coding capacity is a major disadvantage. We might also point out that, for the moment, chipless tag readers most often require specific development, which results in cost. Chipless tags, though rarely marked commercially at present, can be grouped into several subfamilies. We will focus now on the tags that are considered to be the most promising. These are based on:

1) the manufacture of organic transistors via inkjet printing, an approach that remains largely prospective and whose operating principle remains similar to that of chipped tags but with much weaker performance [SUB 05];

2) the acoustic and optic properties of certain materials, an approach that has already been commercialized [HAR 02];

3) the EM response of printed or engraved electronic RF circuits [JAL 05a, JAL 05b, MUK 07a, PRE 08a, PRE 08b, ZHA 06, ZHE 08].

The third point corresponds to the example given in the introduction, and it is this approach that we will now develop. If we go back to the example of the C-shaped tag, a coding technique consisting of shifting a trough within a frequency range was introduced. This principle is repeated for multiple resonators so as to increase the quantity of information on the tag. There are other approaches, which most often enable a compromise between surface coding density and frequential coding density [TED 12]. In order to best understand the chipless solutions currently available, it is important to focus again on coding. Of the various methods of coding binary information, the two simplest approaches to implement consist of (see Figure 4.10):

– noting the presence (or lack thereof) of a distinctive part of a signal (peak or trough for example) in a temporal or frequential range (in the latter case we can make an analogy with OOK modulation in a temporal system); or

– precisely measuring the duration or interval (in the temporal or spectral representation of the signal, respectively) between the presence of distinctive parts of a signal (in the temporal case, we can make an analogy with pulse-position modulation (PPM) modulation, classically applied in temporal systems).

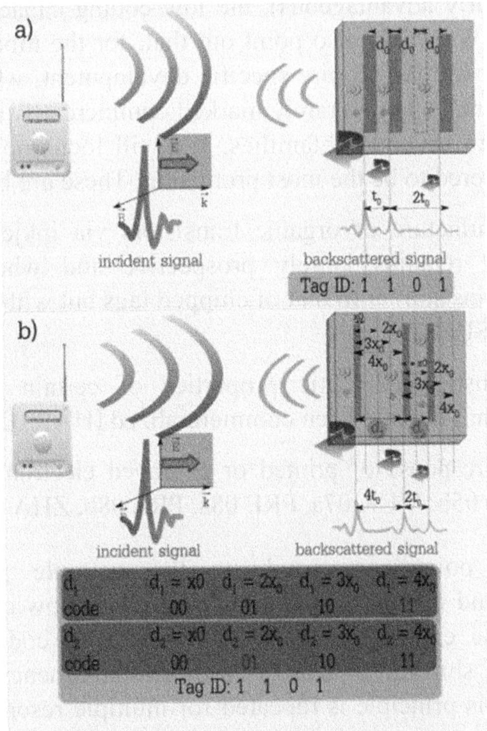

Figure 4.10. *Illustration of two ways of encoding data based on a temporal approach. On the left we can see the pulse sent on the chipless tag located to the right. Here the tag is symbolized by rectangles that act as obstacles to the wave propagating from left to right in the structure. Each obstacle generates a reflection, which gives rise to the reflected signal, thus containing the RF signature of the tag. The two approaches described here use: a) the presence or absence of a specific reflector, and b) the space between the reflectors. In both cases, the information coded corresponds to the same ID, specifically 1101. By replacing the abscissa axis featuring time in this example with frequency, the same coding principle can be used in a frequential approach. For a color version of this figure, see www.iste.co.uk/perret/radio.zip*

Signals are most often EM fields for which we are interested either in the amplitude or the phase. The example described in the last section is based on a frequential approach. In this case, the idea is to use the UWB band so as to be able to have a fairly large band permitted. Thus we can see that this approach makes it possible to have larger quantities of information.

In the frequential domain, it is possible to code information by focusing on variations in frequency of the amplitude of the wave re-emitted toward the reader. This is possible using resonant elements placed near a transmission line [PRE 08a, PRE 08b] or by playing on the resonance frequency of the several dipoles put together [JAL 05a, JAL 05b]. It has also been shown that it is possible and particularly interesting to code information on the phase variation of the wave [MUK 07a, MUK 07b]. This is, in this case, a complex load connected to the antenna is used to modify the phase of the backscattered wave. However, in terms of robustness of communication for example, temporal approaches are generally of greater interest. In this case, the design of such devices is based above all on the concept of reflection between different environments, which can be obtained in the domain of microwaves by the presence of discontinuities or characteristic impedance variations. A rudimentary approach consists of placing a certain number of discontinuities at different distances in order to obtain a signal whose information is coded by the position of impulses, where these discontinuities can be obtained by using localized [ZHA 06] or distributed [ZHE 08] capacities.

4.3.2. *Frequential tag and temporal tag: definition*

We will now go back to what we have just called "frequential tags" and "temporal tags". This is an important distinction and requires additional information. These two concepts have been introduced through examples of chipless tags given in the literature. As shown in Figure 4.11, depending on the approach used, information can be read either from the temporal signal reflected by the tag, or from its spectrum (typically the tag's RCS). It is clear that we can switch from one representation to the other via a Fourier transform, and that in this respect they can be seen as equivalent. It is for this reason that whatever the tag, it can be read with a "temporal" measurement bench (based on example on the sending of a pulse) or a "frequential" measurement bench (based on a VNA, for example). We can see that this aspect of measurement does not discriminate and therefore cannot be used to define two different families of chipless tags. However, if we are interested in a signal from which we can directly (without a mathematical transformation) read a tag's ID (that is, carry out the decoding stage), the latter is necessarily expressed either in frequency or in time (see Figure 4.11). It is based on this observation that we can define what corresponds to the "frequential tag" and "temporal tag" designations.

Now that the principle, the latest developments and the principal definitions of chipless technology have been introduced, it is of interest to go back to traditional identification systems, so as to position this new technology of RF barcodes in relation to existing technologies.

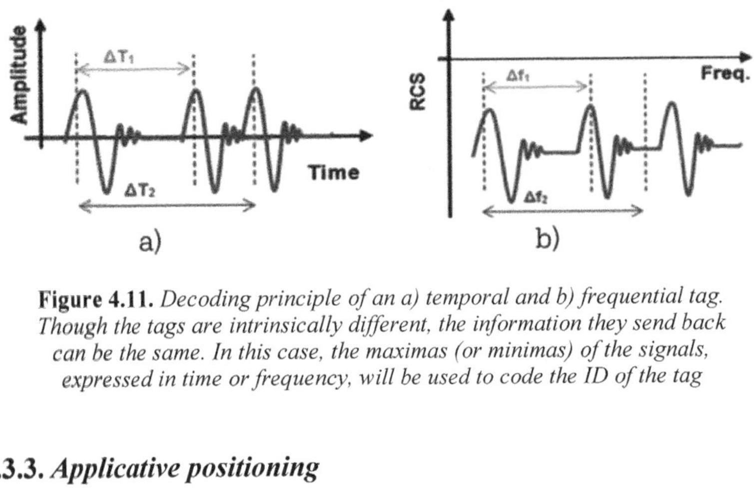

Figure 4.11. *Decoding principle of an a) temporal and b) frequential tag. Though the tags are intrinsically different, the information they send back can be the same. In this case, the maximas (or minimas) of the signals, expressed in time or frequency, will be used to code the ID of the tag*

4.3.3. *Applicative positioning*

In terms of intended applications, we can situate chipless RFID technology midway between classic RFID and barcode technology. This becomes all the more true if we consider criteria such as read range (in the order of one meter), or even simply system complexity. This last point is an important one. Chipless solutions are much less complex than classic RFID, both in terms of hardware (lack of circuit/electronic chip, etc.) and of software (the development of performant communication protocols is not required). This level of simplification and absence of protocol enables rapid identification that is compatible with a large number of industrial requirements.

It is clear that chipless technology has advantages that have ensured the mass deployment of barcodes (low cost, simplicity of application, and reliability in harsh environments) while adding the flexibility of RFID; that is, the use of radio waves for wireless communication. In fact, unlike barcodes, this technology offers the possibility of being read remotely without direct line-of-sight and, in some cases, without a specific orientation between the tag and the reader. Until now, these qualities were proper to RFID technologies. Any obstacle between the tag and the reader will hinder the reading of barcodes. The major disadvantage of this is that human intervention is very often required to position the tag correctly in relation to the reader. Moreover, these limitations are restrictive in terms of the environments in which barcode technology can be used; it is confined to indoor environments where no masses of particles can obstruct the surface of the barcodes and make identification impossible.

Compared to chipped RFID tags, this new family of tags offers better mechanical reliability with regard to the external environment and a unit cost that is compatible with mass utilization. This latter point is a fundamental one, since cost is currently the principal hindrance to the development of classic UHF RFID. Technologically speaking, these tags are compatible with standard printing techniques using conductive ink or, like RFID, with standard printed circuits. If we consider the fact that more than two-thirds of the current price of an RFID tag is composed of the price of the chip and of installing the chip on the antenna, it is clear that, even using current production techniques, the price of chipless tags can be reduced considerably. Additionally, in most cases these tags can be printed directly on the objects to be identified.

SAW technology [HAR 02] is the only chipless technology being distributed commercially at present. Though it is true that this technology does not use a silicon chip and can therefore be considered a chipless solution, it is also true that in terms of application it belongs more to classic RFID than to the chipless approaches being discussed in this book. In fact, it is based on the use of an acousto-optic material (or piezoelectric transducers), which is most often affixed to an antenna to form the tag, as in UHF RFID. Because of this, it is difficult to print on standard materials (paper, cardboard, plastic, etc.) and even harder to print on the object being tagged. Neither is it low in cost, which distinguishes it quite sharply from the chipless approaches we have discussed.

If we replace all of these observations in the projected economic context [IDT 14] described in Chapter 1, we can see that chipless technology is in the process of fulfilling the expectations mentioned. As an example, the application of a barcode costs around \$0.005, whereas an RFID tag still costs around \$0.10 to \$0.30. The projection for 2020 lowers the average price of an RFID tag to \$0.01 (\$0.04 for chipped tags and \$0.004 for chipless RFID tags). However, we will see that it is possible to produce chipless tags with a unit price on the order of around €0.004 (see Chapter 5). This shows that this technology is currently capable of achieving the objectives put forth by forecasters. As an example, the intended applicative sector may particularly concern a specific individual identification (for example for secured applications) or a serial identification (for example the identification of a flow of separate objects filing past the reader) in a closed loop.

The chipless solution is highly relevant in environmental terms as well. Given the very large market for RFID, as previously mentioned, an extremely large number of tags will have to be produced. The recycling of tags, which will number in the hundreds of billions per year, is becoming a major challenge for the future. Chipless tags do not contain chips, which means that they do not contain silicon, and are manufactured using materials with high recycling potential. In fact, tags are both recyclable and potentially liable to provide a traceability function to facilitate the recycling stage. Moreover, these tags will guarantee optimal energy management, since they are passive devices that function via reflection (without batteries or a power supply provided by the reader). The power required to read them is much lower than that of chipped RFID tags as well.

Now that we have put chipless RFID in perspective in terms of positioning, we will end this chapter by emphasizing the key elements of this technology. We will discuss its advantages and, as a family of low-cost (potentially printable) chipless RFID tags, compare different approaches based on these criteria.

4.4. Advantages

Chipless RFID is based principally on a radar approach that is extremely simple to understand. However, though its operating principle is simple, it is also true that its application can prove difficult. This has a great deal to do with the limited capacity to code information on chipless tags. Based on this observation, it seems important to equip ourselves with merit indicators reflecting the performance of an approach, particularly in comparison to another approach.

The need to be able to store an ever-greater quantity of information involves a significant optimization phase for chipless tags, and thus some key parameters must be known. Chipless technology is radically different from classic identification approaches, just as major differences can exist between one family of chipless tags and another. For this reason, it is vital to begin by defining the relevant values to be taken into consideration, if only to help in determining the size of the tags. However, we will see that many of these values are highly dependent on one another. Additionally, it is difficult to improve one value without causing another one to worsen. This leads us to define multiple merit indicators. From there, compromises must be made in terms of design so as to favor this or that criterion over another one. In this section, we will begin by listing the characteristics that are important to take into consideration when designing chipless RFID systems. Next, we will introduce merit factors that will help us to give an informed indication of the tag's performances, and also to compare one approach with another.

4.4.1. *Different ideas to take into consideration*

A chipless tag must carry out multiple functions; specifically, capturing an EM wave generated by a reader; modifying the wave so as to link it with its identifier; and sending the wave back toward the reader. Obviously, the challenge here is to carry out all these functions without an integrated circuit. The fundamental parameters of chipless RFID are the quantity of information, the frequency band used, the dimension of the tag and its robustness of detection in real environments (see Chapter 5).

4.4.1.1. *Quantity of information in chipless tags*

Due to the absence of an electronic circuit, the quantity of information that can be inscribed on a chipless device is a critical point. With the exception of SAW technology [IDT 14], which requires a piezoelectric substrate and consequently is not low cost this quantity of information is generally low, on the order of several dozen bits. Compared to other identification applications this quantity of information is at best similar to what is possible with classical barcodes. However, it is much lower than the quantity possible with classic RFID tags, all the more so because UHF tags can have an internal memory and thus a very high storage capacity. From the user's point of view, this quantity of information is virtually transparent. It may have a slight impact on the price or on the read time needed to recover all of the information stored in the memory. However, it has no effect on the dimensions of the final tag or on the communication frequency. These observations are quite different from the ones that we can make concerning chipless RFID. In this case, the quantity of information will impact many parameters: the tag's dimensions, as well as the frequency band used.

4.4.1.2. *Frequency band*

The choice of frequency band is vital for two reasons. The first reason is based on the fact that, generally speaking, the larger the band used, the more possible it is to have tags with a large coding capacity. This is all the more true when a frequential coding approach is used. In this case, the tag's capacity depends on the number of frequential windows associated with each resonator. The more windows we have, the more different configurations there are. As an example, considering a reader-side frequential resolution in the order of 50 MHz, it is feasible to code approximately a dozen bits per GHz.

The second reason has to do with purely regulatory factors. RFID identification is subject to numerous regulations imposing standards and normalization. Regulatory authorities set frequency bands, transmission power per frequency and maximum communication time between tags and readers. The use of UHF RFID is necessarily limited by these regulations. It is restricted to the frequencies allocated to industrial, scientific or medical applications (ISM, or Industrial–Scientific–Medical bands). These bands have the advantages of being legally free, but do not benefit from international standardization. They are most often too narrow to be used in chipless RFID. A series of International Organization for Standardization (ISO) standards concerning the identification of RFID tags (ISO/CEI 18000) has been developed. Different sections of these standards describe the communications possible at these various authorized radio frequencies. In addition, alongside these international norms, there are "proprietary systems" on the market that use their own communication protocols. Chipless RFID technology does not use any communication protocol; this immediately puts it outside ISO standards, which concern only chipped RFID devices. For this reason, it seems that a simple and effective way to operate in chipless RFID is to comply with UWB norms [ETS 08]. In these conditions, it is possible to cover a large frequency band (several GHz) using a short-impulse signal. Using this method, it is possible to use chipless RFID tags able to contain around 40 bits of information – a quantity equivalent to barcodes (the EAN-13 barcode). The issue of norms in relation to chipless RFID has been addressed in the reference article [VEN 11] and the thesis [VEN 12a]. For all of these reasons, it is of interest to connect the tag's capacity to the frequency band used. This criteria would make it possible to characterize the tag's performance in relation to the usage frequency, which is a very important value, as we have seen. This leads us to the notion of the frequential coding density of chipless RFID tags.

4.4.1.3. *Tag size*

Many chipless applications rely on the use of resonators (frequential approach) or discontinuities (temporal approaches). Each of these elements is the cause of information coding. Let us take the example of resonators. Generally speaking, there is a range within which, when the number of resonators is increased, the coding capacity is increased as well, up to a certain limit. The increase in the number of resonators necessarily results in an increase in the surface area of the tag. We must ensure that there is enough space between the resonators so that they do not disturb one another. Clearly, it is indispensable to link the concept of quantity of information to that of geometric surface area of tags. Thus we speak of the surface coding density of chipless RFID tags.

It is also of interest to define other parameters related to tag performance in terms of use. These can include read range and the sensitivity of tags to their usage environment.

4.4.1.4. *Read range*

Here again, the read range of a chipless tag cannot be defined in the same way as for a passive UHF RFID tag. In the case of chipless RFID, there is no problem of impedance adaptation or of minimum power necessary to activate the chip. Only the tag's capacity to re-radiate enough power to the reader and the sensitivity of the latter need to be considered. This brings us back again to the radar equation, and it is the tag's RCS that will give us an indication of the tag's capacity (or lack thereof) to function at a great distance from the reader [VEN 12a]. In practice, respecting UWB norms, it has been shown that the read range is 50 cm with a frequential approach [VEN 12b]. It can be 2–3 times greater for a temporal approach, but for considerably less information [NAI 13a, NAI 13b].

4.4.1.5. *Sensitivity of tag to the environment*

This last criterion, though rarely taken into consideration, is very important. From the moment we begin to consider a possible industrial application, this technology must be able to be used in a real environment and not just in an anechoic chamber [VEN 12b,

VEN 12c, VEN 13]. This issue of the tag's sensitivity to its environment is all the more pressing if we take the example of frequential chipless approaches. In this case, to increase coding capacity, for a given frequency band, it is interesting to have resonators with a significant quality factor. In this way we can optimize the frequency band by using a large number of these resonators, while associating a part of this band with them. However, the presence of any object near the tag will interact with the latter by modifying the resonance frequencies. The more significant this phenomenon, the more reading will be negatively affected. This example shows us that the tag's capacity is related to its sensitivity to its environment. Thus, in a very restricted environment, it will be necessary to reduce the tag's performance in terms of coding capacity in order to retain a constant read rate. Conversely, a very sensitive tag can no longer be used as an identifier, but rather as a sensor. We will also see in the next chapter that the application of self-compensation techniques makes it possible to improve the tag's detection robustness significantly.

4.5. Conclusion

In conclusion, in this chapter we have introduced the operating principle of chipless RFID, paying particular attention to describing approaches based on the use of potentially printable tags. We have seen that the principle of coding information; that is, the ability to code the tag's identification number, is based on the generation of a specific temporal or frequential signature. This temporal signature can be obtained via the generation of different reflections that are temporally separated. In a frequential system, the information will be readable based on the tag's response spectrum.

The latest developments in basic chipless techniques have been addressed, showing the main advantages and limitations of these approaches. Likewise, we have examined the applicative potential of this technology by situating it in relation to existing technologies. To conclude, a comparison of several approaches has been made through the introduction of merit factors, which we have used to identify the principal limits.

The next chapter focuses on the methods that have been applied to address most of the limitations discussed in this introductory chapter.

4.6. Bibliography

[ETS 08] ETSI, TS 102 754 V1.2.1 – Electromagnetic compatibility and radio spectrum matters (ERM); short range devices (SRD); technical characteristics of detect-and-avoid (DAA) mitigation techniques for SRD equipment using ultra wideband (UWB) technology, 2008. Available at http://www.etsi.org/deliver/etsi_ts/102700_102799/102754/01.02.0160/ts_102754v010201p.pdf.

[FLE 02] FLETCHER R.R., Low-cost electromagnetic tagging: design and implementation Doctor of Philosophy, Massachusetts Institute of Technology, 2002.

[HAR 02] HARTMANN C.S., "A global SAW ID tag with large data capacity", *Proceedings of the IEEE Ultrasonics Symposium*, Munich, Germany, vol. 1, pp. 65–69, 2002.

[IDT 14] IDTECHEX. Available at http://www.idtechex.com/.

[JAL 05a] JALALY I., ROBERTSON I., "Capacitively-tuned split microstrip resonators for RFID barcodes", *35th European Microwave Conference,* Paris, France, pp. 4–6, 2005.

[JAL 05b] JALALY I., ROBERTSON I., "RF barcodes using multiple frequency bands", *IEEE MTT-S International Microwave Symposium Digest*, 12–17 June 2005.

[MET 02] METOIS E., YARIN P.M., SALZMAN N., *et al.*, "FiberFingerprint Identification", *Proceedings of the Third Workshop on Automatic Identification*, Tarrytown, NY, pp. 147–154, 2002.

[MUK 07a] MUKHERJEE S., "Chipless radio frequency identification by remote measurement of complex impedance", *European Microwave Conference*, pp. 1007–1010, 2007.

[MUK 07b] MUKHERJEE S., "Chipless Radio frequency identification (RFID) device", *1st Annual RFID Eurasia Conference*, Istanbul, pp. 1–4, 2007.

[NAI 13a] NAIR R.S., PERRET E., TEDJINI S., "Group delay modulation for pulse position coding based on periodically coupled C-sections", *Annals of Telecommunications*, Special Issue on Chipless RFID, vol. 68, nos. 7–8, pp. 447–457, 2013.

[NAI 13b] NAIR R.S., PERRET E., TEDJINI S., "A temporal multi-frequency encoding technique for chipless RFID based on C-Sections", *Progress in Electromagnetics Research B*, vol. 49, pp. 107–127, 2013.

[NIK 12] NIKITIN P., "Leon Theremin (Lev Termen*)*", *IEEE Antennas and Propagation Magazine*, vol. 54, pp. 252–257, 2012.

[PER 13] PERRET E., HAMDI M., TOURTOLLET G.E.P., *et al.*, "THID, the next step of chipless RFID", *7th IEEE RFID Conference*, Floride, pp. 261–268, 30 April 2013-2 May 2013.

[PRE 08a] PRERADOVIC S., BALBIN I., KARMAKAR N., *et al.*, "Chipless frequency signature based RFID transponders", *European Conference on Wireless Technology (EuWiT'08)*, pp. 302–305, 27–28 October 2008.

[PRE 08b] PRERADOVIC S., BALBIN I., KARMAKAR N.C., *et al.*, "A novel chipless RFID system based on planar multiresonators for barcode replacement", *IEEE International Conference on RFID*, pp. 289–296, 16–17 April 2008.

[PRE 10] PRERADOVIC S., KARMAKAR N.C., "Chipless RFID: bar code of the future", *IEEE Microwave Magazine*, vol. 11, no. 7, pp. 87–97, 2010.

[SUB 05] SUBRAMANIAN V., FRECHET J.M.J., CHANG P.C., *et al.*, "Progress toward development of all-printed RFID tags: materials, processes, and devices", *Proceedings of the IEEE*, vol. 93, no. 7, pp. 1330–1338, 2005.

[TED 12] TEDJINI S., PERRET E., VENA A., *et al.*, "Chipless and Conventional Radio Frequency Identification: Systems for Ubiquitous Tagging", NEMAI CHANDRA K. (ed.), *Mastering the Electromagnetic Signature of Chipless RFID Tags*, pp. 146–174, 2012.

[VEN 11] VENA A., SINGH T., TEDJINI S., *et al.*, "Metallic letter identification based on radar approach", *General Assembly and Scientific Symposium, URSI*, Istanbul, pp. 1–4, 2011.

[VEN 12a] VENA A., Contribution au développement de la technologie RFID sans puce à haute capacité de codage, PhD Thesis, Université de Grenoble, Valence, France, 2012.

[VEN 12b] VENA A., PERRET E., TEDJINI S., "Design of compact and auto compensated single layer chipless RFID tag", *IEEE Transactions on Microwave Theory and Techniques*, vol. 60, no. 9, pp. 2913–2924, 2012.

[VEN 12c] VENA A., PERRET E., TEDJINI S., "A fully printable chipless RFID tag with detuning correction technique", *IEEE Microwave and Wireless Components Letters*, vol. 22, no. 4, pp. 209–211, 2012.

[VEN 13] VENA A., PERRET E., TEDJINI S., "Design rules for chipless RFID tags based on multiple scatterers ", *Annals of telecommunications*, special issue on Chipless RFID, vol. 68, nos. 7–8, pp. 361–374, August 2013.

[WIK 14a] WIKIPEDIA, Leon Theremin. Available at http://en.wikipedia.org/wiki/Leon_Theremin.

[WIK 14b] WIKIPEDIA, The Great Seal bug. Available at http://en.wikipedia.org/wiki/Thing_(listening_device).

[WIK 14c] WIKIPEDIA, Electronic article surveillance. Available at http://en.wikipedia.org/wiki/Electronic_article_surveillance.

[ZHA 06] ZHANG L., RODRIGUEZ S., TENHUNEN H., *et al.*, "An innovative fully printable RFID technology based on high speed time-domain reflections", *High Density Microsystem Design and Packaging and Component Failure Analysis (HDPapos'06)*, 2006.

[ZHE 08] ZHENG L., RODRIGUEZ S., ZHANG L., *et al.*, "Design and implementation of a fully reconfigurable chipless RFID tag using inkjet printing technology", *IEEE International Symposium on Circuits and Systems (ISCAS'08)*, pp. 1524–1527, 18–21 May 2008.

5

Development of Chipless RFID

5.1. Introduction

We saw in the last chapter that chipless radio frequency identification (RFID) technology[1] is quite recent, having flourished significantly since the late 2000s. We will look now at the principal developments in chipless technology that have occurred at LCIS since 2009. The result of this work is a vision of chipless technology intended above all to be practical and will be illustrated through several examples. Unlike earlier explorations in this area, we have always chosen to take the practical aspect of the application of the solution into consideration no matter what the concept introduced. This is why the questions of system cost, frequency bands to be used, development of a reader, and the taking into account of the reading environment are essential points to which we have provided solutions. As stated in the previous chapter, the chipless boom is closely linked to that of barcodes. From this point of view, it will be developed only if its practical application remains as simple as possible. As an example, we will not use discrete components in the tag; using such a component, even a capacitance, does not seem to us to be compatible with the idea of chipless technology. Even though it is by definition "chipless" or silicon-free, it remains a component affixed to the tag, which will have a direct impact on the cost and application of the latter. For this reason, we may legitimately wonder if a chip could not

1 We are not including SAW technology in this term, as its cost and application differ fundamentally from the approach we will be examining.

be more adapted, due to the large number of functionalities it would make possible.

Next, we will retrace the series of ideas leading to the development of chipless tags as they currently exist. This will show that, conceptually speaking, they can be broken down into two categories; one based on the association of antennas and a filter for the coding part of the information, and the other consisting of those introduced at the Laboratory of Conception and Integration of Systems (LCIS), which are based solely on the concept of radar components coding information intrinsically.

In 2006, there were a dozen articles addressing aspects of remote, chipless radio frequency (RF) identification [BAL 09, JAL 05a, JAL 05b, MAN 09, MUK 07, PRE 08, PRE 09, ZHA 06]. All of these articles presented a "circuit" approach to the tag; that is, a device including two separate functionalities (both theoretically in terms of the concept and concretely in terms of the geometry of the tag); specifically, one part to receive and transmit the RF signal (the "antenna" part) and one part where information can be recorded (which is most often represented by a filter) (see Figure 5.1, [PRE 09]). Only the approach introduced by Jalaly *et al.* gave a different reading [JAL 05b]. The original, and clearly precursory, concept of this article consisted of transcribing the principle of the barcode structure into RF; it is for this reason that the term "RF Barcode" was introduced. In this scenario, parallel bars of different widths, like optic barcodes, were used (see Figure 5.2) to produce a chipless tag. The differences here arose from the fact that the patterns were realized with conductive materials; they had different lengths and, of course, their reading did not rely on an imagery approach but rather on the radar principle. In addition, at each geometry (length and width: the width here makes it possible to adjust the quality factor, while the length sets the resonance frequency), a different RF signature is obtained, thus enabling an identifier to be recorded in the tag. In order to make them more recognizable and to distinguish one identifier from another with precision, a ground plane is present on the rear side in order to attain a more resonant structure. The decoding part is thus simplified, since the presence (or absence) of a frequency trough is used to code a "1" (or a "0"). As noted in the article, from an

RF point of view, these structures correspond to a network of microstrip dipoles, each one of which is associated with a filtering capacity[2]. This is the interpretation used subsequently to introduce a variant of this initial structure [JAL 05a], utilizing the concept of antennas loaded by an element (in this case a capacitance) used to code information. These dipole antennas are no longer in short-circuit in this scenario, and they have dimensions that are not different (as in [JAL 05b]) but rather all identical and loaded with capacitances. Each capacitance value is used to modify the resonance frequency, and thus to code the identifier. The principal interest of this lies in the tag manufacturing procedure, which can be compatible with a large-scale production approach. The part that is identical to every tag (the array of dipole antennas) can be produced in large numbers. These blank tags could, for example, be produced with mass production techniques such as flexography. It would then be possible to "write" the identifiers on these tags (in [JAL 05a] this is a matter of soldering capacitances whose values set the desired ID). Even if we abandon the idea of soldering a capacitance, this principle of two-stage production remains highly interesting from a practical point of view. As we will see, flexography makes it possible to obtain tag costs that are much lower than those of classic passive ultra high frequency (UHF) tags. For this stage of ID writing, where the objective is to assign specific characteristics to each tag (which is usually a matter of modifying their geometry), inkjet or laser excision printing techniques are flexible enough to be used at high speeds during the production of chipless tags.

Based on the structures introduced by Jalaly *et al.*, it seems logical to try to reduce the number of antennas, for reasons of surface overload, or to reduce interference, for example, while making the coding part of the information associated with each antenna more complex in order to achieve a significant coding capacity. Thus, rather than having n antennas associated each time with a component coding an ID, structures including two antennas (one for transmission and the other for reception) and a filter composed of n resonators have been

2 "They consist of arrays of microstrip dipole-like structures that behave as resonant bandpass or bandstop filters tuned to predetermined frequencies", from [JAL 05b].

introduced, like the structure shown in Figure 5.1 [PRE 09]. This tag is based on the use of two antennas, a microstrip line, and spiral resonators placed along the line and removing part of the signal at predetermined frequencies linked to the geometry of the spirals. Many variants of this structure have been introduced, in which resonators are arranged along a transmission line. The presence or absence of these resonators directly determines the ID contained in the tag[3].

This has resulted in a highly widespread family of tags whose operating principle is based on the association of antennas (transmission–reception) and a filter to code information. At first glance this approach may seem like a significant improvement, but it has the major disadvantage of having an extremely low quantity of information per unit of surface area. Antenna matching[4] is also very sensitive to the object on which the tag is placed. Generally, this makes these tags unsuitable for practical use. Moreover, from a conceptual viewpoint, this approach breaks completely with the innovative and influential idea of the chipless technology that can be glimpsed in the initial structure of Jalaly's RF barcode. Another conceptual vision of chipless technology is possible, based on the idea that a "unitary object" – or "RF encoding particles", hereafter referred to as REP – can intrinsically have a characteristic radar signature, and can thus be associated directly with an identifier. The reception, signal processing and transmission functions are not separated from one another, both in terms of design and of the geometry of the tag (see Figure 5.3). In this case, the REP act simultaneously as transmission, reception and filtering devices [VEN 11a , VEN 12b]. These tags do not have an antenna or a transmission line, as these components are

3 Note that temporal tags, with a single antenna and a delay line, also exist [SHR 09], while in frequential systems (an antenna with a filter) no tag of this type seems to have been introduced. The reason for this is simple; with two antennas, it is easy to uncouple the reflected signal from the transmitted signal by positioning two antennas with linear polarization perpendicularly to one another. This does result in additional bulk, however, which is most often non-negligible.

4 Due to the large frequency range needed, monopole antennas are often used [NIJ 12, NAI 13b]; that is, antennas based on a partial ground plane, which has a significant impact on antenna matching.

too cumbersome and restrictive for the development of chipless RFID tags.

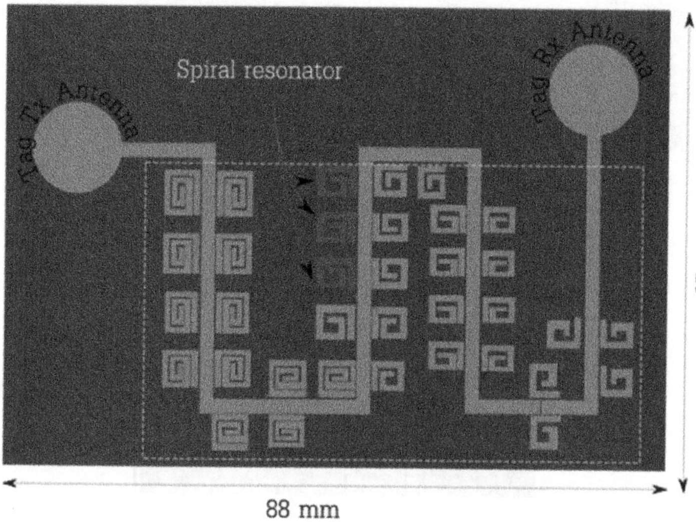

Figure 5.1. *Tag representative of one of the first chipless RFID tags realized. The tag is composed of two clearly distinct parts; one part to receive and transmit the RF wave (Tx and Rx antennas), and the other part used to code information (in this case spiral multiresonators placed along a microstrip line act as an RF filter). For a color version of this figure, see www.iste.co.uk/perret/radio.zip*

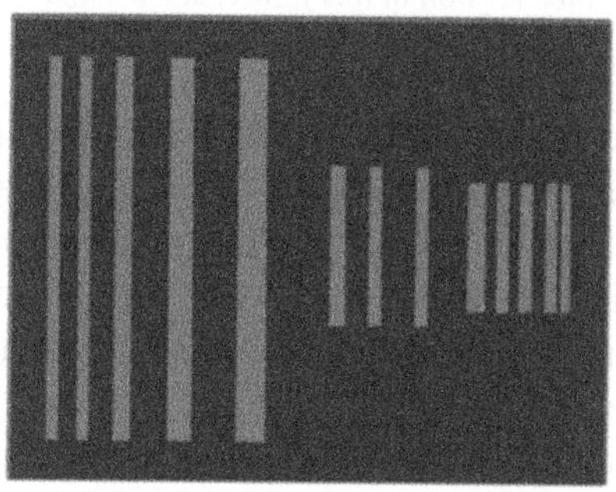

Figure 5.2. *The first "RF barcodes" introduced in 2005 by Jalaly et al. This structure is one of the very first chipless RFID tags with the potential to be compatible with low-cost production and application techniques. Unlike the impression given by the figure above, each dipole has a different length, enabling the coding of one bit*

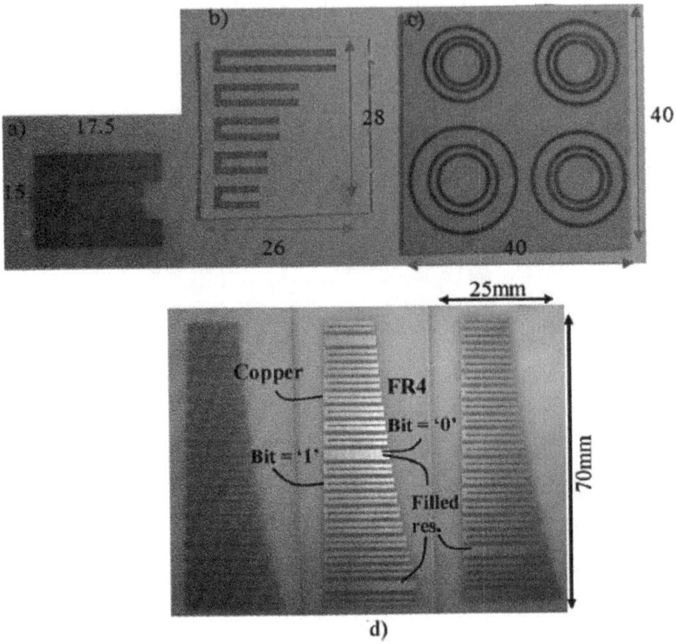

Figure 5.3. *Photograph of REP tags: a) tag in the form of nesting C shapes, showing the highest coding density per surface unit; b) tag produced on paper composed of multiple rectangular SRR based on a hybrid coding technique combining phase and amplitude coding; c) tag in rings, invariant compared to polarization. They have a record coding capacity of 49 bits; d) three configuration of a 20-bit tag functioning with OOK coding. All of these tags are compatible with regulations in effect for RF communications For a color version of this figure, see www.iste.co.uk/perret/radio.zip*

These tags, like RF barcodes, must remain as simple as possible. The surface area associated with the presence of an antenna is too bulky, and the problems linked to the use of this type of approach are extremely constrictive, like the issue of antenna performance (problems of mismatching between the antenna and the line; problems of effectiveness related notably to dielectric losses; problems related to the frequency band, which must cover the operating frequencies of all resonators). It has been shown that the REP approach makes it possible to achieve better coding effectiveness per unit of surface area. In addition, these tags are smaller, and a ground plane is not absolutely necessary. For now, the REP approach is compatible only with frequency coding, while the circuit approach can be used in both frequency [PRE 09] and time [CHA 06, GIR 12]. In the literature we may note two other articles discussing approaches [BLI 09, MCV 06] which present similar nature with the REP. The first introduces a tag with multiresonators based on second-order Peano curves, while the second uses an elliptical dipole with λ/4 notches to code information.

If we return to the REP approach, we can see that this purely "radar" concept, where the wave reflects off the surface of the tag-object, is more efficient in terms of energy, enabling us to dispense with any "circuit" considerations. The absence of a ground plane here is of considerable importance in applicative terms [PER 10, VEN 11a], making the approach totally compatible with printing the tag directly on the object to be tracked.

As described in the previous chapter, for the practical application of chipless technology, it is imperative to have a direct, one-to-one link between the shape of the object and its RF signature. This is used to determine the tag's pattern from the identifier, and vice versa. Coding based on the presence of a peak or a trough at a given frequency remains the simplest approach to implement [VEN 12b]. It appears quite relevant, then, to use resonant objects as RF encoding particles and to produce the tag in that way. Likewise, for obvious reasons of overload, λ/4 resonators are preferable and, if possible, should be positioned very close to one another. These observations have led to the use of certain basic forms [VEN 12a, VEN 13a], some of which are shown in Figure 5.3. Figure 5.3(a) represents the tag we chose in Chapter 4 to introduce the operating principle of chipless technology. Here, the data is coded using a structure composed of partially folded conductive strips, inside which short-circuits are positioned to code the ID [VEN 12b]. The whole device can be printed in one or two stages. In the second case, as previously noted, the precise positioning of the short-circuits in the slots can be done later, in order to record the desired information on the blank tag. Because the popularity of chipless technology is closely linked to the production cost of tags, the REP approach provides a solution that is as simple to implement as barcodes. The possibility of producing these chipless tags on a large scale has been demonstrated [VEN 12a,

VEN 13c, VEN 13d], with a unitary cost of around 0.4 euro cents. Tags on a paper substrate, based on the REP approach, have been produced using flexography, an industrial printing procedure with an extremely high rate of production (see Figure 5.3(b)). To conclude this illustration of REP tags, Figure 5.3(c) shows a tag with a record capacity of 49 bits [VEN 12d]. This shows for the first time that chipless technology can contain as much information as the well-known EAN-13 barcode (on the order of 41 bits).

The rest of this chapter is organized around the introduction of new designs and improvements introduced since 2009, focusing particularly on the following points:

– increase of the coding capacity and density of chipless RFID tags;

– improvement of communication between tags and the reader;

– practical application of chipless RFID technology, in tandem with:

- the modeling of tags' behavior in order to obtain their geometry for a given identifier (and the solution of the inverse problem);

- reduction of tag costs;

- design of tags compatible with the permitted frequency bands;

- realization of a reader for chipless tags.

We will begin by discussing what may be considered as one of the basic issues of chipless RFID technology: its storage capacity for recording the identifier.

5.2. Coding capacity and density of chipless RFID tags

The coding capacity of chipless RFID tags has been at the heart of the chipless issue. For more detail, different methods of coding information in chipless technology are described in [TED 12]. This obstacle remains a hindrance to the large-scale development of this technology. The reality is that its coding capacity remains low compared to that of competing technologies. This is why, a great deal of work has been done in recent years to increase this capacity considerably, resulting in the record number of 49 bits [VEN 12d]; the previous record was 35 bits [PRE 09]. The essential point has been to show that it was possible to break the symbolic 41-bit barrier, which represents the total quantity of information EAN-13 barcodes can contain (of the EAN-13 barcodes used most frequently today). To achieve this result, it was necessary to conduct a significant amount of work focusing on various technical and regulatory factors. Coding capacity also depends on: (1) the intrinsic performance of the resonators used, (2) the technique used to code the information and (3) the development of a reader which, among other aspects, must comply with effective transmission and reception norms.

5.2.1. *Performances of resonant patterns*

A detailed study focusing on a sample of resonant structures has been conducted; the results are presented in [VEN 12a] and [VEN 11a], and the performances of each of these are compared in [VEN 11a]. Each resonant pattern is characterized by a bandwidth, a quality factor, and its radar cross-section (RCS) level at resonance. The higher the quality factor, the more resonators can be associated with a given bandwidth, and thus the quantity of information is increased. However, certain parameters other than the performance of the pattern in itself will also come into play; the dynamic of the reader, its frequential resolution or simply the nature of the substrate on which the pattern is imposed will necessarily have an impact on the system. The choice of pattern, then, results from a compromise between multiple factors. The use (or lack of use) of a ground plane will also change the situation considerably. We most often seek a high RCS level, again for the sake of reading performance (an increase of the signal-to-noise ratio results – at constant transmission power – in a greater read range). It is also necessary to consider the issue of pattern overload; the coding techniques used are usually based on the use of multiple resonators with the same shape but different dimensions, positioned side-by-side, as shown in [VEN 11a]. The effects of coupling between different patterns must be taken into account through the updating of design rules in order to minimize them and to ensure that each pattern remains as independent as possible from its neighbors.

With regard to the frequency band used, which is limited, and given that these devices are big consumers from this point of view, it is crucial to use the band as effectively as possible; that is, to optimize the use of allowed frequencies. In this approach, based on the use of multiple resonators, each of which is associated with a frequential slot, it is clear that it is the pattern and not the legislation that will most often limit the frequency band. Indeed, working in the UWB band (3.1–10.6 GHz), it appears in this case that the scatterer high order frequency modes limit information quantity; the scatterer with the lowest resonance frequency may have a higher resonance inside the UWB band, which limits the usable frequency band. In the end, the simple C-shape shown in Figure 5.3(b) has proven to be a good compromise. This shape has been used in various configurations and is notably compatible with the implementation of advanced coding techniques, which are discussed in the section 5.2.2.

5.2.2. *Information coding techniques*

Because of the recent history of chipless RFID technology, certain aspects related to information coding performance have been little studied. Only basic on–off keying (OOK) coding in frequency systems and pulse-position modulation (PPM) coding in temporal systems (see preceding chapter) have really been used. This point is vital, however, since it enables the optimization of the allocated frequency band and has a direct impact on surface and frequential coding density, as we saw in the example described in Chapter 4. This is why the coding of information in chipless technology has been the subject of particular study in recent years; a classification of the different methods possible is given in [TED 12, VEN 12a]. This most interesting approach from this perspective has been to attempt to associate several bits of information to a single resonator in order to reduce surface coding density. This goes back in a sense to transcribing the PPM modulation method into a frequential system (see Figure 4.10 of Chapter 4). This is known as a coding approach by frequency position. It has been shown that it is possible to encode several bits per resonator [VEN 11c]. A second advance has been to demonstrate that it is possible to impact several parameters independent of signal for the same coding particle [VEN 11a]. C-shaped patterns have made it possible to couple classic coding on signal amplitude with phase coding of this same signal. Playing upon several key parameters of the pattern, it has been shown that it is possible to control the frequency position of amplitude peaks, independent of phase jumps.

This hybrid coding has increased the number of configurations possible. To represent these various states, the concept of the constellation diagram has been introduced [VEN 11a]. This is similar to what is used to represent signal modulation for digital transmissions; the difference here is that we are not situating ourselves in time, but rather in frequency. Other original types of coding have been introduced, such as the one presented in [VEN 12e], for which it has been demonstrated that it is possible to differentiate between signals by working with the polarization of the wave.

5.2.3. *Transmission and reception standards*

This point has already been largely addressed in the previous chapter, therefore we will limit ourselves here to contextualizing the issue of choice of frequencies within the development of chipless RFID technology. The first studies conducted on this point were done as part of the PhD work of Arnaud Vena [VEN 12a]; up to that point, the issue of the practical application of chipless technology had not been addressed. Articles mentioning read range did not absolutely fall within the context of regulations. At best, the ranges evoked then were higher than what it is possible to obtain; at worst, the frequency bands used were not actually allowed for a usage of this type.

As previously noted, chipless technology falls outside of ISO standard pertaining to RFID. Regulations allow either the use of industrial scientific and medical (ISM) free bands or compatibility with UWB norms [ETS 08]. We are particularly interested in UWB norms; however, tags compatible with certain ISM frequencies have also been developed [NAI 12, VEN 12d]. The main interest of approaches based on ISM frequencies lies in the fact that the maximum power authorized remains very high compared to what it is possible to do in UWB. For example, 2 W effective radiated power (ERP) can be sent in Europe in the 865.5–867.6 MHz band (equivalent to 4 W effective isotropic radiated power (EIRP) in the United States in the 902–928 MHz band). Likewise, for the ISM band between 2.4 and 2.4835 GHz, the European Telecommunications Standards Institute (ETSI) allows an EIRP power of 10 mW (10 dBm) for generic use. A signal of 25 mW EIRP can even be transmitted in 150 MHz of the ISM band at 5.8 GHz. With regard to UWB, the maximum spectral power density (SPD) authorized is only -41.3 dBm/MHz, but it enables us to obtain the best coding capacities with chipless technology by playing on the cyclical ratio and using a band of several GHz. Moreover, read ranges greater than 2.5 m can theoretically be obtained, a figure achieved using a pulse respecting the UWB limitations and tags with an RCS of -30 dBsm [VEN 12a]. For slightly less optimistic tag performance values (typically on the order of -50 dBsm), read ranges of 50 cm have been observed in both theory and practice.

Note that SAW technology uses an ISM band (2.45 GHz/10 mW EIRP) and makes it possible to have read ranges of several meters; this is comparable to classic UHF RFID [HAR 07]. We can see through this example that for much lower transmission powers (10 mW EIRP instead of 4 W EIRP in the best-case scenario in classic RFID), chipless technology, and particularly SAW, offers far better performance in terms of read range. In this, chipless clearly consumes less energy. This can be explained mainly by the fact that no energy is lost in powering the chip. This allows for a quasi-optimum management of energy resources. A detailed theoretical study presenting the power balances and maximum read ranges possible depending on the approaches used (notably UWB) is given in [VEN 12a].

To conclude this section on coding capacity, we will reiterate that a great deal of work has been done on this point. We have also seen that quantity of information is at the heart of the chipless issue, as it affects many other factors. Though there have been marked advances, notably demonstrating that it is possible to code as much information in a chipless tag as in an EAN-13 barcode, new techniques that are different from those currently used in this field are expected. In reality the coding capacity of chipless tags remains low for numerous applications. It is clear that the market for tags containing small amounts of information is still extremely limited. We may estimate that in order to penetrate the mass market, the memory capacity of a tag must be at least 128 bits [IDT 14]; this number corresponds to the quantity of information necessary to be able to implement an Electronic Product Code (EPC), which is the benchmark global standard.

5.3. Improvement of the robustness of detection of chipless RFID tags

It is clearly important to have a significant information capacity, but is it still necessary to be able to read the information recorded on the tag easily? At first glance, this is not necessarily easy, for the following reasons:

– the signal sent back by the tag is extremely weak. For application, it is preferable to have tags with small dimensions, which means weak RCS. For example, a usable signal is generally much weaker (up to 40 dB at least) than the one sent back by the object on which the tag is placed [VEN 13b];

– this object will have another impact as well; its electromagnetic (EM) properties will actually modify the tag's signature. If, for example, we use PPM coding in frequency (thus focusing on the position of resonances inside specific frequential windows), a shift in frequency (related to the properties of a high-permittivity material, for example) may occur, rendering the tag unreadable in the best-case scenario. At worst, it may send back erroneous information. A material with losses will smooth the resonances somewhat, also leading to the loss of the information embedded in the tag. It is worth noting that this issue occurs more often with tags that do not have a ground plane [VEN 13b];

– the last point we will mention concerns problems of polarization, which are manifested here by the necessity of orienting the tag in relation to the reader. Chipless tags like the C-shaped tag will operate for a given polarization; the challenge here is to be able to produce tags that do not vary with polarization, so as to simplify the reading method and to establish a clear distinction from barcodes.

As we will see, these difficulties are not insurmountable and solutions do exist.

The first order of business is to take one or more calibration measurements in order to avoid as much as possible being impacted by disturbances related to objects located near the tag we are attempting to read. The ideal here is to be able to measure the environment without the tag (but with the object being identified, which we will henceforth write as OBJECT; to avoid confusion, this measurement is written as $S_{21}^{isolation}$), and then with a reference tag on the OBJECT (S_{21}^{ref}). The last measurement in this case is that of the tag on the OBJECT; that is, in its environment of use (S_{21}^{tag}). As it happens, the reference tag may simply be a metal plate; that is, a reflective structure whose RCS is known. Note that in general, the OBJECT we are attempting to identify is theoretically unknown. Based on these three measurements and using expression [5.1], it is possible to obtain very good results:

$$\sigma^{tag} = \left[\frac{S_{21}^{tag} - S_{21}^{isolation}}{S_{21}^{ref} - S_{21}^{isolation}}\right]^2 . \sigma^{ref} \qquad [5.1]$$

where σ^{tag} is the RCS of the tag thus obtained by measurement, and σ^{ref} is the RCS of the reference object that must be known to it. The reference object may simply be a square conductive plate a few centimeters long on each side. Its RCS value in frequency can be obtained either by using classic analytical formulas or by conducting an EM simulation.

Using this approach, we can see that the results of measurements and simulations are very consistent. We can also see, as shown in [VEN 11b], that this approach is compatible with both a frequential measurement (classically carried out in the laboratory with the aid of a vector network analyzer (VNA)) and a temporal measurement (taken with a pulse and an oscilloscope) [VEN 13b]. The problem here is a practical one. It is difficult to imagine an application where it is possible to remove the tag from the OBJECT being tagged and replacing it with another element, while still carrying out the three measurements. Only the empty measurement – that is, of the environment without the OBJECT and the tag – is easily accessible from a practical point of view. We might even imagine retaking this measurement at regular intervals in order to take into account any modification of the environment as time passes. The measurement of the environment with the OBJECT but without the tag, however, is much harder to obtain. This is even more true when we consider that the OBJECT is unknown. This is the same problem that exists for the measurement of the OBJECT with the reference. It is important to note that by using the calibration approach summarized in expression [5.1], it is required to know implicitly the object on which the tag is placed, as well as the position of the tag relative to the reader. This last point is very important and this issue generates very little comment in the academic press. Indeed, for any movement of the tagged object, it is essential to repeat the three measurements. Also, with this approach, we do not have a real reading area as in traditional RFID, which makes it almost useless in practice. Thus we can see the limits of this approach, and the need to use specific methods to address this problem. It is also clear that, unlike the classic RFID approach, in this case any temporal modulation of the tag (called Modulated Scattering Technique, or MST approaches) becomes impossible.

We will now move on to results related to the question of improving reading in a real usage environment. We will begin with the case of REP frequential tags. For these tags, various techniques have been applied, such as the principles of depolarizing tags and tags that are invariant by polarization, and finally coding that compensates for the frequential gaps caused by the dielectric nature of the object we are seeking to identify.

5.3.1. *REP approach (frequency domain)*

5.3.1.1. *Principle of wave depolarization*

The technique implemented has consisted of seeking a way of increasing the power of the usable signal both in relation to the noise level and, especially, in comparison to all of the reflected signals (in this case the part of the signal that does not contain any useful information). As we have noted, the usable signal is very weak, and one way of recovering it more effectively is to attempt to isolate it. The approach resulting from this observation is to attempt to put a second transmission "channel" in place. There are several ways to do this. The first way consists of changing the frequency of the reflected useful signal; for this, a nonlinear component of the tag can be used. However, besides the very low power conversion rates that can result, the addition of secondary components would go against the vision of chipless technology, which must prioritize simplicity of implementation and low cost above all. Another solution consists of playing on the polarization of the wave; part of the reflected wave (the usable part) by the tag must be orthogonal to the wave sent. The part of the wave reflected in the same direction as the sent wave is not used. To execute this type of function in the case of circuit-type chipless RFID (that is, a tag including two antennas and a transmission line like the one shown in Figure 5.1), it is sufficient to orient one of the tag's antennas at a 90° angle to the other. Note that only the signal propagating along the transmission line – that is, the usable signal – will be depolarized. In reality, this is more or less true; it is above all a function of the shape of antenna used that is able to send back a wave in cross-polarization. This operating principle is based on the observation that in practice, few everyday objects will depolarize the RF wave from 90° by themselves. This is the case for dielectric objects. For those that are conductive, there will always be time to "secure" the reading zone if a problem is encountered; thus only the useful component is depolarized, which makes it possible to obtain the functionality sought. To give a concrete example we must only take a cardboard box filled with sheets of paper; the latter sends back a component in cross-polarization i.e. 25–30 dB lower than the one in co-polarization [VEN 13b]. The same observations can be made concerning the human body [AND 14].

The question now becomes one of attempting to transpose this principle into the context of the REP approach. Along with the intuitive description given here, a theoretical demonstration has been carried out in order to validate the approach [VEN 13b], and different families of tags based on this principle have been developed (see Figure 5.4). At the design level we have tried to favor resonances where a significant portion of the associated currents circulate perpendicularly to the direction of the incident wave. This type of current is shown in Figure 5.5. It clearly appears that half of the current is oriented at 90° in relation to the source of excitation, which will enable the backscattering of a component in cross-polarization. The RCS relative to these coding particles is also shown in Figure 5.5. We can see that it is possible to have peaks in the response in cross-polarization with high signal levels (on the order of -23 dBsm). A considerable quality factor is obtained here by coupling several elements with one another (two for the tag shown in Figure 5.4(a), and five for the one shown in Figure 5.4(b)), and by using a rear ground plane.

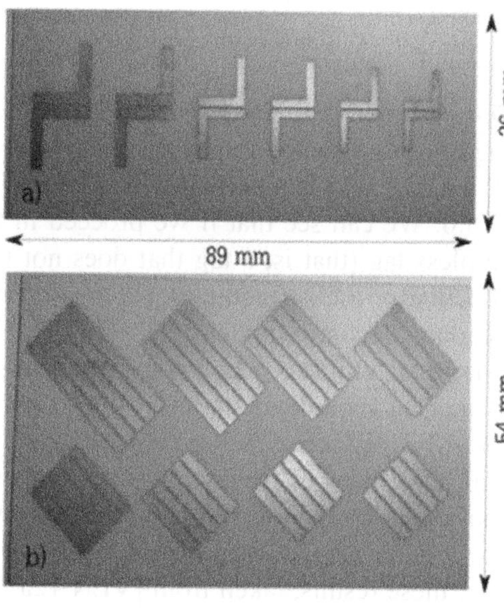

Figure 5.4. *Photograph of two "depolarizing" tags: a) dual-L resonators and b) shorted dipoles oriented at 45°. These tags operate with a ground plane*

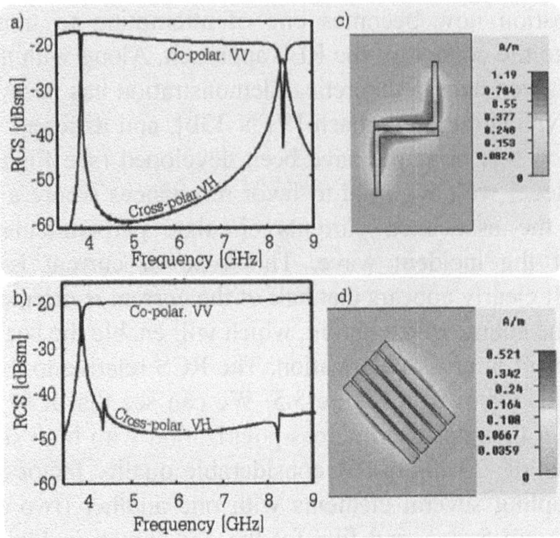

Figure 5.5. *RCS of coding particles in co-polar (VV) and cross-polarization (VH): a) dual-L resonators, b) shorted dipoles oriented at 45°. c)–d) Surface current density at resonance of the depolarizing coding particles shown in Figure 5.5. For a color version of this figure, see www.iste.co.uk/perret/radio.zip*

It has been demonstrated that it is possible to read these tags using only an empty measurement (without the OBJECT) and, of course, the measurement of the tag on the OBJECT. The principal results are shown in Figure 5.6. We can see that if we proceed in the same way with a classic chipless tag (that is, a tag that does not have the wave depolarization function or a C-shaped tag), the results are completely different, and reading becomes impossible. This result shows real progress in the field of chipless technology; we can see that it is possible to read the tag regardless of its position in relation to the reader (in this case for distances of between 10 and 20 cm) having taken only one measurement of the empty environment. These tags have been read both on a cardboard box filled with sheets of paper and on a pack of water bottles, as well as on metal cans or a human body. To our knowledge, these results, taken from [VEN 12a, VEN 13b] are the first on the subject. Even for "circuit" chipless tags like those shown in Figure 5.1, no characterization of this type has been presented.

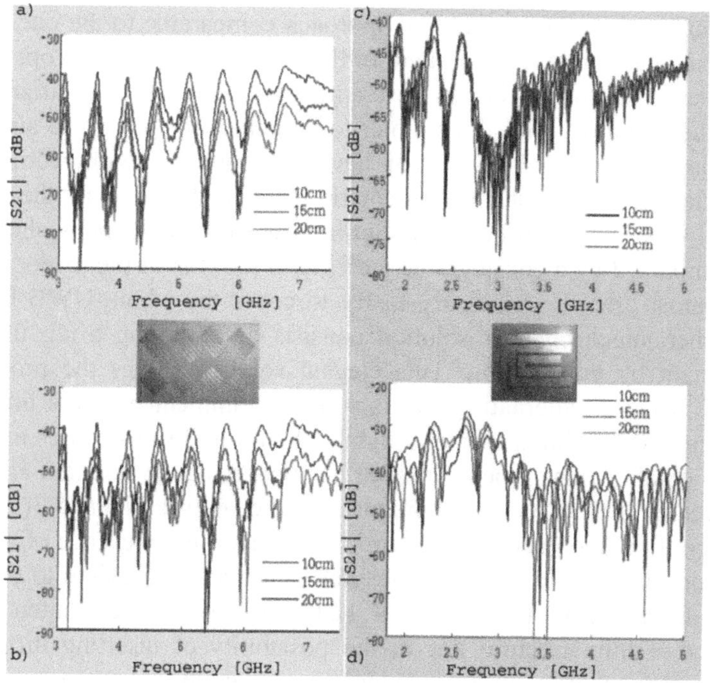

Figure 5.6. *a)–b) Measurements in cross-polarization of depolarizing tags (see Figure 5.4(b)) positioned on a cardboard box filled with sheets of paper, for different distances. c)–d) Measurements in co-polarization of C-shaped tag, also positioned on the same box. In a) and c) the measurements taken with the tag are subtracted from those of the box without the tag, for each position of the box. Conversely, in b) and d) the tag measurements are subtracted from the empty environment value; that is, without the box. For a color version of this figure, see www.iste.co.uk/perret/radio.zip*

5.3.1.2. *Invariant tag in relation to incident wave polarization*

If we take the example of barcodes, we can see that the problem linked to the need to orient the code in relation to the reader (barcodes are not invariant by rotation, even though forms of this type were patented when this technology was introduced) has been partly resolved by making the reader part more complex. This approach can also be applied to chipless RFID, since the vast majority of tags are not invariant by rotation, which results in a signature that fluctuates according to the tag's orientation in relation to the reader's antenna. An EM signature different from the one expected means that the information cannot be read. An approach comparable to the one used for barcodes can also be applied. In this case, it is possible to opt as in classic RFID, for readers that use antennas with circular polarization. Another solution, which is simpler to implement compared to already existing tags, would consist of having an antenna with rectilinear polarization, for which the orientation of the direction of the field could vary in a controlled way. But here as well, these approaches are accompanied by a considerable increase in the cost of the reader, as it will most often be necessary to try to cover the whole UWB band. Another, much cheaper solution consists of designing a tag that is invariant by polarization. This elegant solution solves the problem while using an alternative that is simple to implement. Note that, in this case, the coding particles will take the form of rings, which allows us to obtain resonant behavior, but also invariance by rotation around the center of the structure in the plane of the tag (see Figure 5.3(c)). In this configuration resonance is in $\lambda/2$; however, the presence of a rear ground plane and a low-loss substrate will make it possible to obtain very high quality factors [VEN 12d]. The second very interesting aspect of this structure lies in the possibility of inserting different rings into each other in order to increase the number of resonators, and thus to obtain a very high number of bits, while retaining a very compact surface area. Consequently, the coding principle is simple: making sure to separate the rings sufficiently from one another (i.e. playing on their respective radii), it is possible to make them independent. Each ring will thus have a frequential window attributed to it, in which any variation of the peak inside this window (geometrically, this will correspond to a value different from the radius of the ring) can be used to code information. The nesting of rings naturally results in different radii for each of them, and thus in different resonance frequencies. This means enormous gains in density. The simplicity of the structure enables the use of an analytical equation linking the resonance frequency of each ring with its geometry so that we can directly establish the one-on-one relationship between the geometry of the tag and its ID. In order to use the largest part of the UWB band and thus to increase coding capacity significantly, a version including 12 resonators divided into four packets has been realized (see Figure 5.3(c)). It is essential in this case to separate rings that have radii very close to one another; in this way we retain the independence of the rings while making the most of the very high selectivity of these structures in order to optimize the frequency band available to us. For a total dimension of 4 × 4 cm² the tag can code 49 bits, which is a record in the discipline. Because of the very high selectivity of these resonators, it is possible to associate them with frequential windows of 30 MHz and thus to obtain a very large number of different configurations with 12 independent resonators.

The principal measurement results are shown in Figures 5.7(a) and 5.7(b). We can see the RCS as well as the group delay for 4 tag configurations. The very high selectivity of the resonators can also be seen. The presence here of a ground plane leads to the coding of information on the position of troughs and not that of peaks, as for previous structures (there is a phenomenon of signals in phase opposition, as explained in [VEN 13a]). It is interesting to examine group delay, given that the information is present here as well; we might even note that decoding based on group delay is preferable, since it is easier to distinguish the troughs present here compared to those of RCS. An example is shown in Figure 5.7(c) and 5.7(d), in which we can see the response of the tag including 12 resonators positioned on a cardboard box filled with paper. For frequencies higher than 7 GHz, the troughs on RCS become less and less clear-cut, which can result in decoding errors. The troughs on group delay, however, even though they are weaker in amplitude, remain perfectly usable. These results confirm other observations made in the literature, indicating that phase remains a more robust means of recovering information than amplitude. Note, however, that this duplication of information may be used in practice by the reader in order to increase the read rate. It is also clear in the example of the measurement taken on the cardboard box filled with sheets of paper that the presence of the ground plane provides a screen, eliminating any effects linked to the OBJECT. This is an interesting point; however, again for practical reasons related mainly to the cost of the tag (increased complexity of manufacturing procedures, increased quantity of conductive ink, inability to print the tag directly on the OBJECT), solutions without a ground plane are preferred. For this reason, we expect the OBJECT to modify the tag's response and thus to disturb its identification. It is preferable, therefore, to implement a so-called compensation technique to enable tag reading in the absence of a ground plane. This point is the subject of section 5.3.1.3.

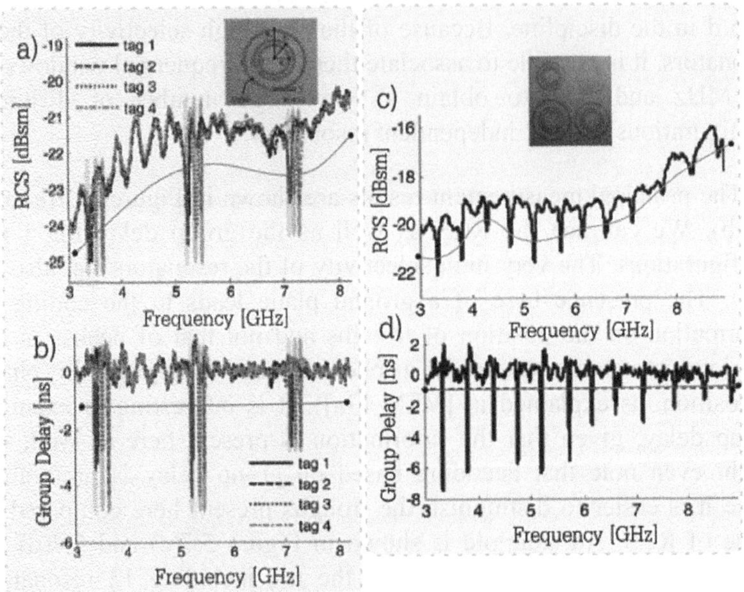

Figure 5.7. *Results of measurements obtained on invariant tags in relation to polarization. a) and b) RCS and group delay, respectively, of 4 tag configurations containing 3 resonators; c) and d) RCS and group delay, respectively, of tag containing 12 resonators*

5.3.1.3. *Increase of reading robustness – auto-compensation techniques*

As we have just noted, the use of tags without ground planes is expected for reasons of price. In this case, the resonators making up the tag will be in direct interaction with the OBJECT being identified, which will necessarily cause a deformation of the tag's signature. This deformation is inherent in the presence and nature of the OBJECT which, again, is assumed to be unknown. Thus, compared to an approach that must remain potentially unprintable at low cost, it is impossible to try to limit this phenomenon. The only available solution is to adapt as far as possible, and this can be done since the deformations induced can be easy to understand overall. If we focus on the case of the dielectric materials of which many objects are made, the deformation observed can be approximated with a translation in frequency of the signal's spectrum [VEN 12b]. This translation is dependent on the effective permittivity of the surrounding environment. It is true that the act of considering a family of objects is limiting in itself, but this approach is linked to the use of RF waves and not to the application of chipless technology in itself. We have seen in the previous chapters of this book that a similar approach was recommended and implemented in the case of UHF RFID. Note also that the vast majority of packaging materials are made of cardboard, which makes it possible to separate materials from the tag that are less favorable from an EM point of view.

As soon as we consider that the deformations expected are generally translations of the spectrum of the tag's signature, we can address this problem simply in terms of the decoding part. In reality, the recovery of the ID from the peaks, or the troughs of the frequential signal, is done in relation to their relative frequency position; that is, in relation to one another, rather than absolutely. An approach of this type has been used successfully both in cases of OOK coding [VEN 12c] and frequency PPM coding [VEN 12b]. It is advisable, then, to distinguish between the two types of coding. We will start with the OOK approach.

5.3.1.3.1. OOK coding

Classically, in OOK coding it is the presence or absence of a peak (or trough) at a given frequency that codes a 0 or a 1. If we apply the auto-compensation method previously introduced, it is sufficient to note the alternation of presences or absences of peaks, for example, to access the information. This alternation must keep the same periodicity in frequency (the frequency delta between two resonators is constant overall), whatever the nature of the dielectric on which the tag is placed. The only condition we must pay special attention to (and which, as we will see, differs from PPM coding in frequency) is related to the processing of resonators whose resonance frequencies fall at the farthest ends of the band (resonators with the maximal and minimal frequencies, respectively). For certain configurations, that is, certain IDs, it is possible for one of these two resonators or even both to be absent, which may distort decoding if no precautions are taken. We may then interpret the absence of a resonator as a frequency shift of the whole signal spectrum, which will lead to a read error. The simplest way to remedy this problem is to impose the presence of these two resonators. In this case they will act as markers indicating the start (lowest resonance frequency) and end (highest frequency) of the useful signal. Note that, theoretically, it would be sufficient to force only one of the two resonators to be present. If we consider a regular (or known) spacing of resonator frequencies, decoding can take place based on a single marker. However, forcing the presence of both increases the robustness of detection, as we can more easily access the information when spacings are modified, for example due to the presence of a material whose effective permittivity is not constant in frequency. An analogy might be made with the frames exchanged in communication networks; we can say that these two markers act as frame "preambles" (for the synchronization of clocks) and of the Frame Control Sequence (FCS) (for the integrity control aspect).

The tags used for this study are simple C-shaped tags (see Figure 5.3(d)) [VEN 12c]. Twenty resonators are used to realize the tag, which, with this type of coding, can be used to code a maximum of 20 bits (1 bit per resonator). Blocking two resonators reduces coding capacity enough to increase detection robustness significantly. The principal measurement results of the study are given in Figure 5.8(a), where the frequential shift of each resonator is given for two different materials (dielectric plate made of Polytetrafluoroethylene (PTFE) ($\varepsilon_r = 2.1$) and Plexiglas ($\varepsilon_r = 3$), with dimensions of 10×10 cm^2 and a thickness of 1.5 mm). Figures 5.8(b) and (c) show the measurements of a tag in free space as well as atop a cardboard box filled with sheets of paper. We can very clearly see the presence of a frequency shift, which distorts the decoding process. A third curve, where the auto-compensation procedure is implemented, is also presented. We can see that it superimposes itself onto the curve obtained in free space, which validates the approach taken.

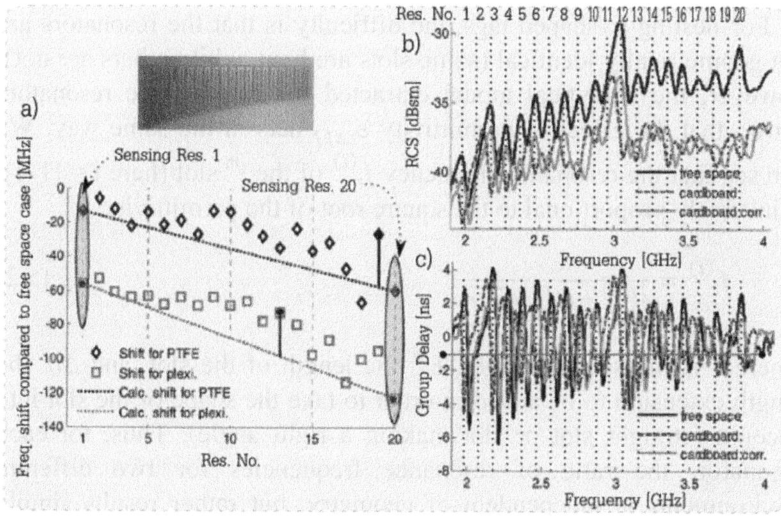

Figure 5.8. *Result of measurements of 20-bit tag used with the auto-compensation technique. a) Resonance frequency of each resonator for two supports (PTFE and Plexiglas). b)–c) RCS and group delay of a 20 resonator tag: comparison between measurements in free space, on a cardboard box filled with sheets of paper, and obtained via auto-compensation (cardboard corr.) For a color version of this figure, see www.iste.co.uk/perret/radio.zip*

5.3.1.3.2. PPM frequency coding

This approach is similar, but the difference here is that a single resonator will code multiple bits. In addition, the frequential interval between different peaks will not be constant, and it is the determination of the value of this interval that will enable us to access the information recorded in the tag. A detailed study of the application of this type of approach is discussed in [VEN 12b]. It involves nesting C-shaped tags like the ones shown in Figure 5.3(a). When we look closely at this tag, we can see that one of its 4 resonance frequencies (specifically the 3rd, i = 3) is unusable for coding. In fact, as we saw in Chapter 4, if we try to modify it, it will cause two other frequencies to shift, which is extremely damaging. It seems quite pertinent to keep it fixed, and to use it to compensate for the frequential shift caused by the presence of the OBJECT. This is why the third resonance has been used as a reference, so as to increase the robustness of detection of the tag.

For nesting C-shaped tags, the difficulty is that the resonators are not geometrically identical (some slots are bent, while others are not); however, the analytical model extracted for each of the resonators shows that the effective permittivity ε_{reff} acts in the same way. We can see that the resonance frequency $f_r^{(i)}$ of the ith slot (here i = [1-4]) is inversely proportional to the square root of the permittivity:

$$f_r^{(i)} = \frac{c}{4\left(L_s{}^i + \Delta L^i\right)\sqrt{\varepsilon_{reff}}} \qquad [5.2]$$

where c is the speed of light, $L_s{}^i$ the length of the slot, and ΔL^i the length extension to be added in order to take the shape of the slot into account (straight slot or slot making a right angle). Thus, for each resonator, the ratio of resonance frequencies for two different environments is independent of geometry, but rather results simply from the permittivities describing each environment:

$$\frac{f_r^{(i)-mes}}{f_r^{(i)}} = \sqrt{\frac{\varepsilon_{reff}^{(i)-mes}}{\varepsilon_{reff}}} \qquad [5.3]$$

where $f_r^{(i)-mes}$ is the resonance frequency of the ith slot measured when the tag is placed on the OBJECT. $\varepsilon_{reff}^{(i)-mes}$ is the corresponding permittivity, meaning that it is linked to the unknown environment near the tag. In [5.3], for a given environment, with the square root of the ratio of permittivities being constant for all of the i resonators, we can link all of the resonance frequencies to one another. For i = [1,2,4], we obtain the following expression:

$$f_r^{(i)} = \frac{f_r^{(i)-mes} \cdot f_r^{(3)}}{f_r^{(3)-mes}} \qquad [5.4]$$

Expression [5.4] allows us easily to determine the resonance frequency value in free space. We obtain this frequency by multiplying the one measured as $f_r^{(i)-mes}$ (tag placed on the OBJECT) by the ratio between the frequency in free space of the fixed resonator $f_r^{(3)}$ (reference resonator i = 3, whose frequency is known), divided by the frequency of the same resonator in the presence of the

OBJECT $f_r^{(3)-mes}$ (that is, obtained by measurement). A parametric study conducted in simulation makes it possible to estimate precisely the errors made as well as the limits of the approach. We can observe an error on the order of 2% for permittivities up to 7, and this for up to 5 GHz. This error allows us to specify the trust interval that can be associated with each resonator; a frequential interval of 100 or 200 MHz is required depending on the resonator in order to code the information without error. The limits of the approach can also be seen in [VEN 12b]. Generally speaking, we may say that it can be used from the moment distinctive signs (peaks or troughs) remain visible. It has been shown that for considerable thicknesses of high-permittivity dielectrics, or even for lossy OBJECTS, these signs will diminish and make any reading – and thus decoding – impossible.

To finish with this point, let us specify, however, that this method is based essentially on the processing of the signal reflected by the tag and not on the design of the tag itself, which facilitates its application enormously. Its area of validity remains sizable, as it considerably increases robustness of reading. We might even add that it is vital for practical applications. The only disadvantage lies in the fact that it will mobilize some resonators, which reduces the quantity of information. However, through the two examples described here, we can see that these reductions are small (two bits in one case and zero in the other) compared to the improvements contributed.

5.3.2. *Temporal approaches*

The principle of the temporal approach was described in the previous chapter; it is simply one of the two coding approaches used in chipless RFID. For this reason it is of interest to undertake research on this theme so that we can compare the two approaches to one another. In this section, we will examine the temporal approach from a reading robustness perspective. As we will see, the main interest of this approach lies in the fact that it can be used to obtain much higher read ranges than the frequential approach. The most telling example concerns SAW tags, which, as we have stated, can be read in real conditions of use with ranges of several meters. However, the principal disadvantage of approaches without piezoelectric materials is the quantity of information that can be coded, which remains extremely low (around 3 bits). With the exception of SAW tags, we note also that little research has been conducted on this theme. Based on these observations, it seems advisable to research techniques to increase the quantity of information in these tags while retaining large read ranges.

The first piece of significant research involving the study of the C-section used to obtain dispersive devices for which the evolution in frequency of group delay could be forced [GUP 10] showed the possibility of using these structures for chipless applications. A circuit-type chipless approach, in which the tag would be composed of two antennas, each connected to a C-section, was used to develop a new family of chipless tags [NAI 13a]. We will now discuss the principal results obtained in terms of reading robustness.

5.3.2.1. *Operating principle*

The tag studied is based on the operating principle of C-sections, which can be seen as extremely compact delay lines whose dispersive character enables increased coding capacity. A C-section line is shown in Figure 5.9(a), which contains a group of 3 C-sections cascaded with a group of 10 C-sections. Besides the C-section structure, the chipless tag is composed of two wide-band antennas, each of which can serve for transmission and/or reception (depending on the tag's orientation in relation to the reader's antenna). As shown in Figure 5.9(b), the two antennas are positioned perpendicularly to one another in order to ensure better isolation of the return signal.

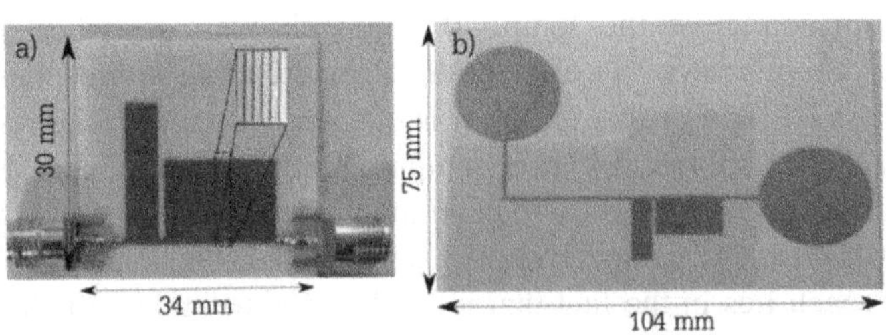

Figure 5.9. *Example of a structure composed of multiple groups of C-sections; that is, of a line including several groups (2 in this case) of 3 and 10 C-sections, respectively. a) Configuration with SMA ports, b) multi-group of C-sections chipless tag*

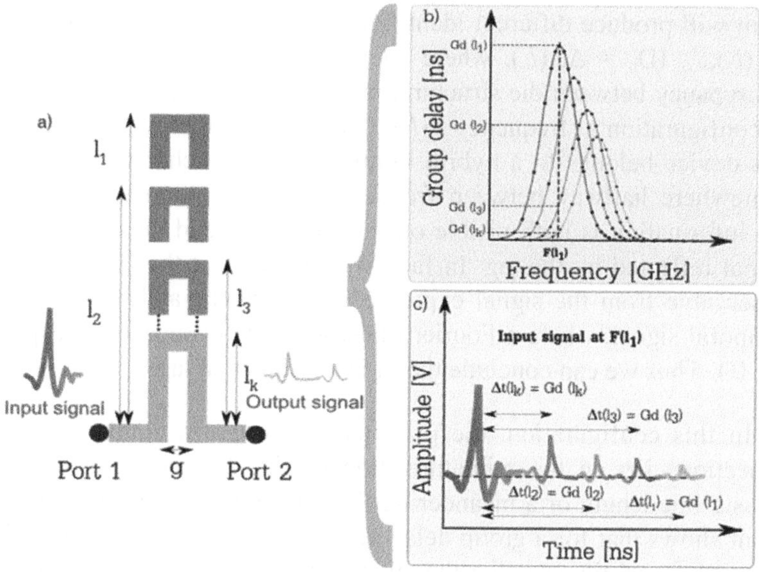

Figure 5.10. *Operating principle of a tag based on the use of a C-section. a) schema of a C-section whose length l_i is variable. b) Group delay depending on frequency for different line lengths l_i. c) Correspondence in the temporal domain at frequency $F(l_1)$. For a color version of this figure, see www.iste.co.uk/perret/radio.zip*

A C-section of length l_i produces a maximum delay for each odd multiple length of $\lambda_g/4$, where λ_g is the guide wavelength (see Figures 5.10(a) and (b)). We can approximate the frequency for which this maximum occurs using the expression: $F(l_1) \approx c/(4^* \, l_1 \, (\varepsilon_{reff})^{1/2})$ where ε_{reff} is the effective permittivity of the microstrip line and c is the speed of light. Any other line length l_i, i=2, 3…n will generate a different delay, lower than l_1 at frequency $F(l_1)$ (see Figure 5.10(c)). From there, when the wave transmitted by the reader arrives at the tag, part of it is directly reflected by the object (we will speak of structural mode), while another part is captured by the antenna and propagates along the C-section, eventually to be re-radiated by the second antenna. Because the duration of the propagation of the wave across the C-section is dependent on the length l_i, the reader will recover two signals (structural mode and antenna mode) separated in time. This time difference is used to code the ID number of the tag. Thus, if we consider n tags including the lengths l_1, l_2... l_n respectively, each of them will produce different identifiers, which can be written as $ID_1 = \Delta t\ (l_1),..., ID_n = \Delta t\ (l_n)$, where $\Delta t\ (l_i)$ is the value of the temporal discrepancy between the structural mode and the antenna mode of the ith configuration at frequency $F(l_1)$. Observe here that we can say that this device belongs to a hybrid family of tags which can be situated somewhere halfway between frequential and temporal tags. Indeed, the information is both visible on the frequency and on the temporal signal reflected by the tag. In fact, we can see that the ID is directly observable from the signal expressed in frequency and not from the temporal signal, where a Fourier transform at F() is needed to capture the ID. Thus we can conclude that it is a frequential tag.

In this configuration, the principal interest of a single group of C-sections lies in the miniaturization of the device compared to a classic line length or a meandering line. In [NAI 13a], a study of this point shows that for a group delay value of 1 ns, the C-section device has a surface 6 time smaller than that of a meandering device, and that the total length of the line when unfolded is almost two times shorter than the meandering line or the straight line. It also shows that for dimensions of around $15 \times 30\ mm^2$, it is possible to obtain very long delays, on the order of 10 ns.

5.3.2.2. *Hybrid temporal-frequential coding and C-section multigroups*

These lines also have another advantage, which will prove to be very important as we continue. The versatile character of these structures in terms of frequency have actually made it possible to produce a new family of chipless tags; specifically, hybrid time–frequency coding. If we consider Figure 5.10(b), we can see the highly dispersive character of the structure. Note the two different types of zones: (1) a zone in which the group delay is large and (2) a zone where it is constant and small. This distinction is even more marked when we increase the number of C-sections or reduce the value of the spacing *g* between lines. Based on this, it is possible to add other groups of C-sections, as shown in Figure 5.11[5]. We will then try to

5 We use the term 'group' to designate a set of *n* C-sections of the same length; it is possible to cascade several groups of different lengths alongside one another.

position the group delay peaks from each group of the C-sections we have just added in the zone where the group delay was previously small and constant (see Figures 5.11(a) and (b)). This leaves us with a structure that behaves in a very unique way: the group delay peaks are in this case independent from one another, which will enable information to be coded on each of them; any variation in the line length of one of the groups will result in a different group delay for the frequency corresponding to the group (see Figure 5.11(b)). This gives us a different configuration than we might recognize with certainty. Moreover, in this case we have a direct, one-to-one link between the evolution of the group delay and the geometry of the tag. Figure 5.12 helps us to visualize the operating principle from a temporal perspective. We can clearly see that the coding here is present in both time and frequency domain representations.

To recover all of the information from the signal contained in the tag, we must filter the temporal signal for each frequency $F(l_{ij})$, with i designating the number of the C-section group making up the tag (see Figure 5.12(a)), with $1 \leq i \leq N$ and N is the number of groups used. The index j indicates, as before, the line length used (with the maximal value having the index 1). Based on this filtered signal, by marking the temporal interval between the structural mode and the tag mode (that is, the two maximums appearing on the envelope of the filtered signal), we obtain the value $\Delta t\,(l_{ij})$. By filtering the backscattered signal N times at frequencies $F(l_{ij})$ $(1 \leq i \leq N)$, we recover the $\Delta t\,(l_{ij})$ corresponding to each group of C-sections, and thus the ID of the tag. These temporal variations lead directly back to the geometry of the tag; that is, the set of n lengths l_{ij} for each group of C-sections (here n variations of length are allowed for each group, or $1 \leq j \leq n$). This example shows the direct link that exists between the geometry and the signature of the tag; we can cause an ID to correspond to each geometry. If we count all of the configurations allowed, we obtain the figure of n^N. This takes us to n configurations in the case of a tag including a group of C-sections at n^N when we cascade N groups. This result is a significant increase of the quantity of information we are able to store.

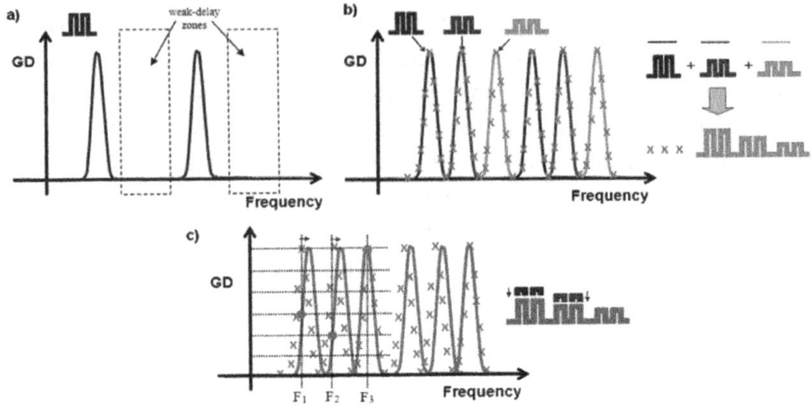

Figure 5.11. *Frequential illustration of the operating principle of the system based on the use of multiple groups of C-sections. Group delay depending on frequency for different configurations: a) a group of C-sections where we can see the appearance of the zones with a very weak group delay, b) placement in series of 3 groups of C-sections of differing lengths. The resulting group delay (cross) is the superimposition of the group delays of each isolated group of C-sections. c) The lengths of the first two groups are reduced in comparison to the initial configuration shown in b). These reductions result in a translation of the parts of the curve that correspond to the first two groups (symbolized by arrows). We obtain one configuration among the 6^3 possible, where 6 and 3 correspond to the number of variations of lengths permitted and to the number of groups, respectively. For a color version of this figure, see www.iste.co.uk/perret/radio.zip*

Simulation curves demonstrating the principle of the cascading of C-sections are shown in Figure 5.13. We can see here, in the frequential domain, an example of 4 different configurations, for n = N = 2.

If we take the definition we have given of chipless tags, specifically, the distinction we have made between temporal and frequential tags, we can see that despite the use of a "line with delay" structure, properly speaking, these multigroup tags should be grouped among frequential tags. It is obviously in the representation of the group delay in frequency that we can see the tag's code, as shown in Figures 5.11(c) and 5.13. The latter is not directly present on the temporal signal; a filtering stage must occur before accessing the ID. However, the fact that we can also extract the signal without shifting into a frequential representation (it is even simpler if the measurement has been taken with a pulse to remain in a temporal representation; in that case we must only filter the signal in time for different frequencies, as shown in Figure 5.12) gives this tag a true hybrid character.

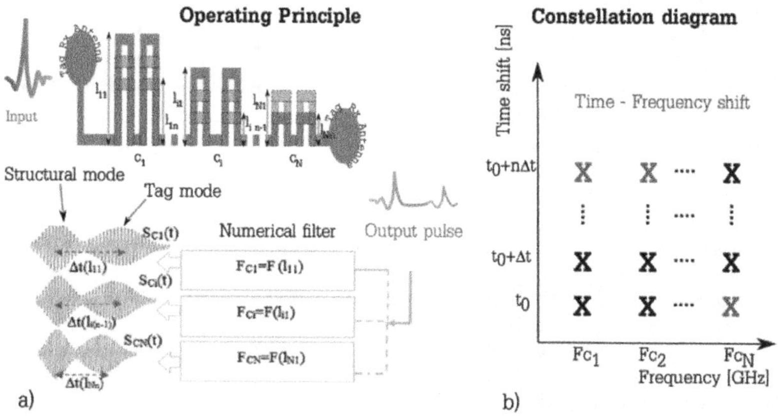

Figure 5.12. *Temporal illustration of the operating principle of the system based on the use of multiple groups of C-sections. a) Operating principle in a temporal representation. In the illustration, each group is made up of two C-sections. b) Frequency-time constellation diagram, coding principle used. For a color version of this figure, see www.iste.co.uk/perret/radio.zip*

Thus we can see that it is possible to increase the quantity of information significantly. However, the difficulty lies in the "selectivity" of devices. The principle schema shown in Figure 5.11 may seem very dissimilar from the simulation curves shown in Figure 5.13. In the second case, areas where the group delay is weak are greatly reduced. In reality, the planar C-sections shown in Figure 5.9 are difficult to put in cascade; it is possible to link two of them at most [NAI 13b], which limits the interest of this type of approach. To increase the selectivity of the structure, it is advisable to reduce the spacing between lines, which increases coupling. However, there are limitations to this in terms of practical application (it is difficult to have gaps between lines that are smaller than 100 μm). Another solution consists of turning to multilayer structures where coupling would be significantly stronger.

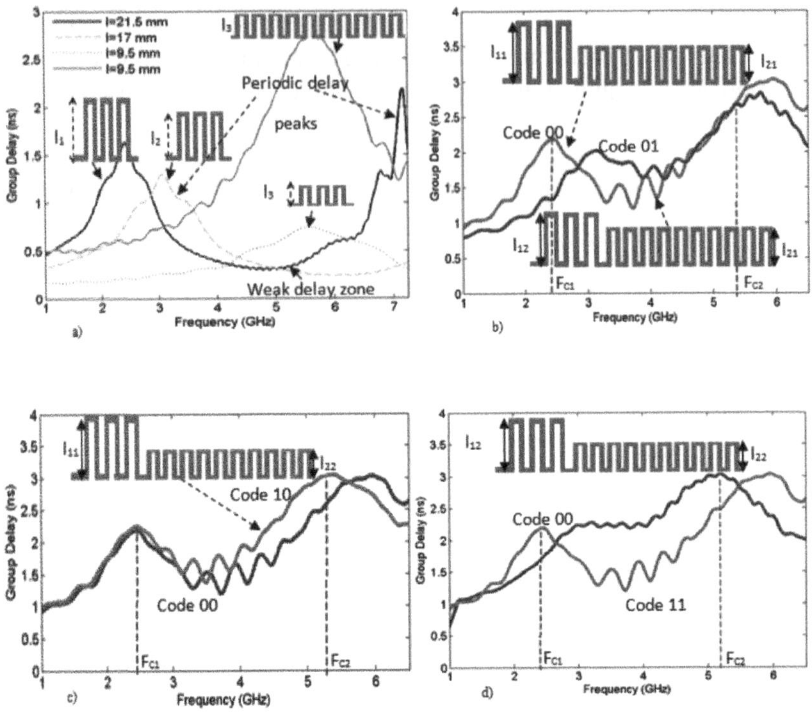

Figure 5.13. *Principle of the cascading of different groups of C-sections. Group delay depending on frequency for different configurations. a) Simulation of the impact of the length and the number of C-sections on group delay. b) Cascading of two groups, configurations 00 and 01, c) 00 and 10 and d) 00 and 11. Each time, only one of the two lengths varies*

In [NAI 13a], it has been shown that the use of multilayer C-section structures can resolve this issue. Figure 5.14 shows a simulation – measurement comparison for a C-section of this type. We can see clearly that selectivity is very high, which will enable us to cascade several groups with one another. With this approach we can code from 6 to 10 bits of information (according to whether we consider between one and three C-section groups) solely in the 100 MHz of the ISM band, around 2.45 MHz. These figures are significant; they are much higher than the 3 bits that can be found in the literature, and it is impossible to obtain so much information on such a small band for classic frequential tags. The coding density per unit of frequency is very much greater in this case than in any of the other structures we have seen. Placing ourselves along the whole UWB band, it seems possible to attain around 42 bits, which rivals what is possible with frequential approaches. In terms of coding, the main interest is in being able to code information on the level of the group delay and not simply on the presence of a peak or trough for a given frequency. This particular characteristic (and not the selective character of the structure) explains the ability to have large quantities of information for very narrow frequency bands. If we do the same thing with the frequential approach introduced at the start of this chapter, until now no structure has been able to code on the magnitude of the RCS; decoding has been done based solely on the position (PPM coding) or presence (OOK coding) of a determining element such as a trough or a peak.

Based on this observation, it is advisable now to examine what has appeared until now to be the main strength of temporal approaches–robustness of detection.

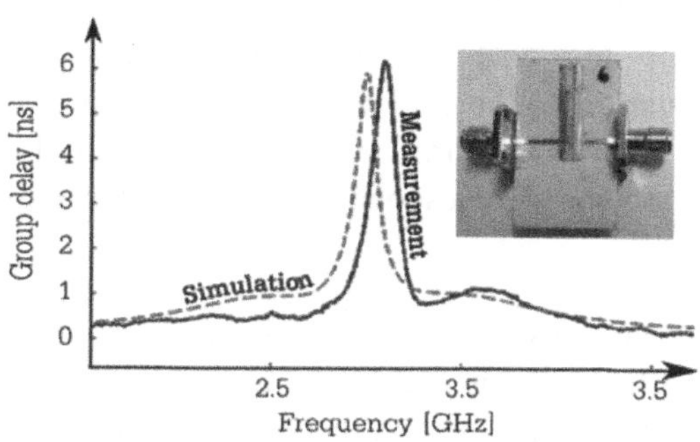

Figure 5.14. *Multilayer C-section. Comparison of group delay in frequency between simulation and measurement*

5.3.2.3. *Robustness of detection of temporal approaches*

The temporal approach is quite interesting from the point of view of robustness of detection. Here, most often, it is enough to be able to note the delay between two successive "echoes". As we have seen, the basic principle is very simple: the temporal approach is given by the structural mode, while the identifier is linked to the time taken by the signal to travel along a line contained in the tag. An example of a tag of this type, composed of a single wide-band antenna connected to a delay line of variable length, is presented in [HU 10, GIR 12]. This example shows that the recovery of the identifier is simple and does not require any advanced calibration procedures, as is often the case in frequential systems. This is an important point, as we have previously explained; it can be prohibitive from an applicative point of view. We can see, effectively, that the C-section based tag can be read directly [NAI 13c]; that is, without using the reference tag. It is true, however, that this tag operates in cross-polarization, but this is not the only explanation. If we look at the literature, the fact of not needing a reference tag is also true for tags operating in simple polarization, like the one discussed in [GIR 12]. This robustness of detection observed in practice is not something easy to formalize from an analytical point of view, but we may say that this is because in the examples presented, there are only two peaks to note rather than a large number of them, which is the case in a frequential system. For this reason, we can see that a reference object is no longer vital to “normalize”, in a sense, the signal received. It is somewhat different for a frequential tag, for which the fact of not compensating for frequency deformations inherent to the external environment (presence of cables, antennas, etc.) can cause large peaks to occur in the band being observed that do not correspond to the resonators, resulting in decoding errors.

Other ideas are possible as well. To explore these, imagine a scenario in which the tag is excited by a pulse. We might argue that for a classic temporal tag, the non-dispersive behavior of the structure makes it so that the two peaks expected (on the temporal signal) are the result of the reflection of all of the frequencies; all of the power, whatever the frequency, contributes to the appearance of these peaks. Conversely, for a frequential tag, each component serves to code different information. We can conclude, then, that the power density per distinctive elements (troughs–peaks) is necessarily lower in this second case. This observation is not related to the fact that more information is coded in a frequential system, but rather to the approach in itself, more precisely the technique based on coding in terms of delay. This observation would remain true even if we could produce a very high number of different delays (imagine being able to do away with problems related to the long lossy line lengths present in the tag). Thus, we would have a very large number of different delays, and each time all of the components of the pulse would contribute to the formation of these two recognizable elements, in this case peaks. However, this observation is only true when the tag's ID is coded on two reflections. If multiple discontinuities are introduced to play on multiple peaks, we find ourselves faced once again with a case similar to the frequential approach.

Note also that it is always preferable to code information on a relative value (such as delay in the temporal approach) rather than on an absolute value (frequency, as with the classic frequential approach). This is also the idea based on which, in a frequential system, the idea has been developed of relative coding linked to the presence of a fixed – i.e. known – resonator, not used to encode data. This eliminates many of the disruptions connected to the measurement environment (see section 5.3.1.3). In the temporal approach there is no need to fix an element; the structure mode is inherent to the object and will thus serve as a reference. Additionally, all of the variations (linear, etc.) that will not impact the delay in the tag will compensate for themselves. Note, however, that the presence of a dielectric material in contact with the tag will still disrupt acquisition, as it causes the delay coding the information to vary. In this case, it would be necessary to add a line with a fixed length, like the one used for the frequential approach, which would enable auto-compensation. In this case we would have three peaks to identify, which also indicates a distribution of the power on the two peaks related to the delay, and thus a compromise to be made.

Finally, still with regard to the idea of a reading system based on the sending of pulses, we may note that, since the ID of a temporal tag is directly present in the backscattered signal, there is no need to use signal processing approaches (time-to-frequency-domain conversion, for example). The decoding part is also very simple, as the useful information is not harmed by digital post-processing errors.

With regard to the robustness of detection aspect, the second notable point has to do with the read ranges that are possible with temporal approaches compared to frequential chipless tags. The most telling example of this is SAW technology, which achieves read ranges comparable to those obtained in UHF RFID. What we will now turn our attention to concerns regulations. As we noted in section 5.3.3 pertaining to norms, it is preferable to be able to use ISM bands, which offer a much higher transmission power to UWB regulation (from 10 to 25 mW on the 100 and 150 MHz of the ISM bands of 2.45 and 5.8 GHz, respectively). Unlike the frequential approach, which requires a large frequency band, temporal approaches can position themselves on one of these bands. For example, SAW systems use mono-pulses in transmission that are centered at 2.45 GHz. For a given distance, the increase in transmission power results in a better detection of the tag (increase of the signal-to-noise ratio). This is why we can associate the issue of regulation with the detection robustness aspect; the use of temporal chipless tags provides us with an undeniable advantage in this regard.

5.4. Practical application of chipless RFID technology

Many things have been noted already on this point, given that many of the objectives targeted were aimed at proving the chipless concept. We will now present the latest aspects that have led to the realization of a portable demonstrator, able to read the tags produced in a real usage environment.

5.4.1. *Design of chipless RFID tags compatible with regulations*

Generally speaking, and in view of regulations, there are three possibilities:

1) working on a narrow frequency band, specifically one of the ISM bands;

2) working on several ISM bands (typically the ISM 2.45 and 5.8 GHz bands);

3) working in UWB (3.1 GHz – 10.6 GHz).

Compared to what has been described previously we have turned toward UWB regulation [ETS 08], principally in order to be able to introduce tags that are simple to produce and include a substantial amount of information.

5.4.2. *Cost of tags*

As we saw when the supporting figures were given, this aspect is essential in chipless RFID. Compared to classic RFID, it is the principal argument in favor of chipless. Cost of tags is therefore closely linked to the future success of this technology. The search for extremely low-cost printing solutions was at the heart of the THID project, financed by the French National Research Agency (ANR) (2010–2013). One partner, the Pulp and Paper Research & Technical Centre (CTP), was tasked with demonstrating that it was possible to produce these tags with printing techniques used in the paper industry; that is, at a very high rate and at very low cost. In order to obtain tags with a cost around the same as that of barcodes, it is vital that they be entirely printed. For this, every localized element reported is prohibited. Likewise, to reduce cost, the use of classic RF materials is not advisable. The materials composing the tags must be low cost but also compatible with classic large-scale production techniques. Another of the challenges here was to manufacture these structures using publicly available dielectric materials such as paper or polymers, which lend themselves to relatively short-term future standardization. In summary, the desire is that the only difference between these tags and barcodes will lie in the use of conductive ink that would be the source of the specific EM signature of the tag. Note also that the fact of using renewable and recyclable cellulose-based materials as coding supports is an undeniable benefit. This work has necessitated the taking into consideration of limitations on potentially usable materials, as well as on the procedures that would be involved in the design, manufacture, and testing of tags. A specific study was therefore made of printing technologies using special inks to produce the conductive parts of the tags, and flexography was the solution chosen [VEN 13d]. This procedure enables the deposit of thick (around 2 to 5 μm), precise ink films at high speeds (see Figure 5.15). This satisfies the printing criteria of electronic components and is therefore compatible with low-cost production, while yielding good conductivity (on the order of 3×10^5 S/m). Moreover, it can be paired with inkjet printing, which is very flexible and allows for the customization necessary in the second phase. A qualitative analysis of the test patterns was conducted and showed that it is possible to reproduce arrays of lines with a minimum width of 50 μm and a spacing of 200 μm without a risk of short-circuiting.

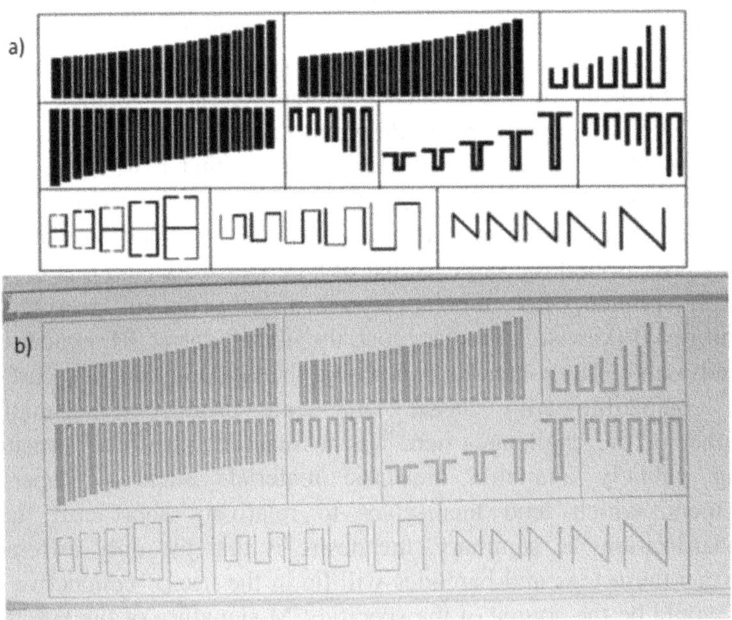

Figure 5.15. *Examples of printings produced using flexography on paper materials with conductive ink: a) snapshot used containing 10 tags, b) printout obtained. In this example, the total area of the print zone is 30 × 10 cm²*

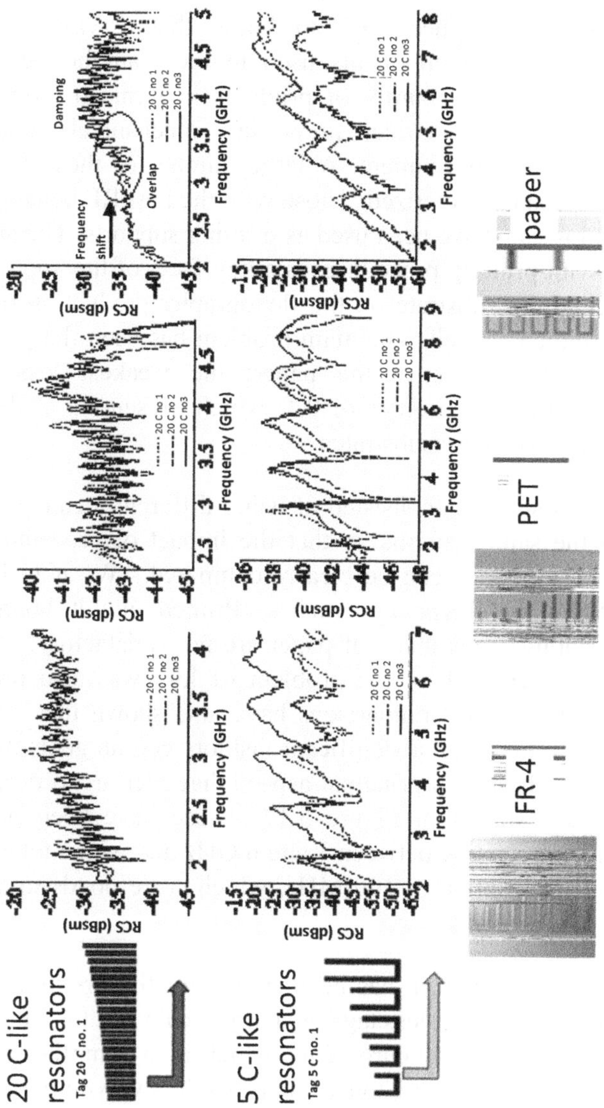

Figure 5.16. *Comparisons of RCS measurements of chipless tags produced in different ways based on identical patterns. Left to right: classic 1-layer FR4 printed circuit board, production by inkjet catalyst printing on PET, and printing via flexography on paper. Top line: tag including 20 resonators; bottom line: tag including 5 resonators*

At the same time, the RF characteristics of paper materials have been realized. It is necessary to know that a vast number of the procedures and components used in the manufacture of paper generate an enormous family of products in terms of composition, surface properties, electrical properties, mechanical content and sensitivity to the environment. A large number of these families of paper have been characterized. Those with the best RF behaviors – i.e. the fewest losses – have been used as printing supports. The affinity of materials with printing procedures is a key facet of the approach, and the impact of the substrate on the flexography production procedure has been studied as well The main problem has been the presence of very significant losses in the paper; the weakest loss tangents encountered are on the order of 0.1, which presents a challenge for producing tags on these substrates.

The initial state has consisted of using different means to produce tags with the same patterns so that the impact of procedure on the resulting performance can be clearly compared. We note in Figure 5.16 that, unlike classic approaches (Printed circuit board, inkjet catalyst printing), tags made of paper are not satisfactory. A specific design phase centered on the problem of loss was also undertaken [PER 13, VEN 13d]. Comparisons have also shown that in order to increase print quality in a significant fashion, certain geometries must be avoided. These two principal aspects have led us to design the 6 resonator tags shown in Figure 5.17. These have been printed on glazed paper, operating between 3 and 6 GHz and with a total capacity of 15.5 bits (resolution of 100 MHz, which corresponds to a 0.5 mm variation of the gap *g*).

Based on this study, it has been shown for the first time that it is possible to produce chipless tags on a large scale and for a unit cost on the order of around 0.40 euro cents, which is in compliance with the 2019 projections noted previously. This result is very important as it is 20 times cheaper than the UHF RFID tags currently available on the commercial market. This figure goes some way toward proving the considerable potential of the approach.

5.4.3. *Production of a reader for chipless technology*

The last point we will mention here concerns the reading part of the approach. In order to fully validate the applicative potential of this solution, it is vital to commit ourselves to perfecting a reader capable of reading the tags. This reader must have a reasonable cost (similar to that of UHF RFID readers, meaning lower than €1,000) and must be transportable.

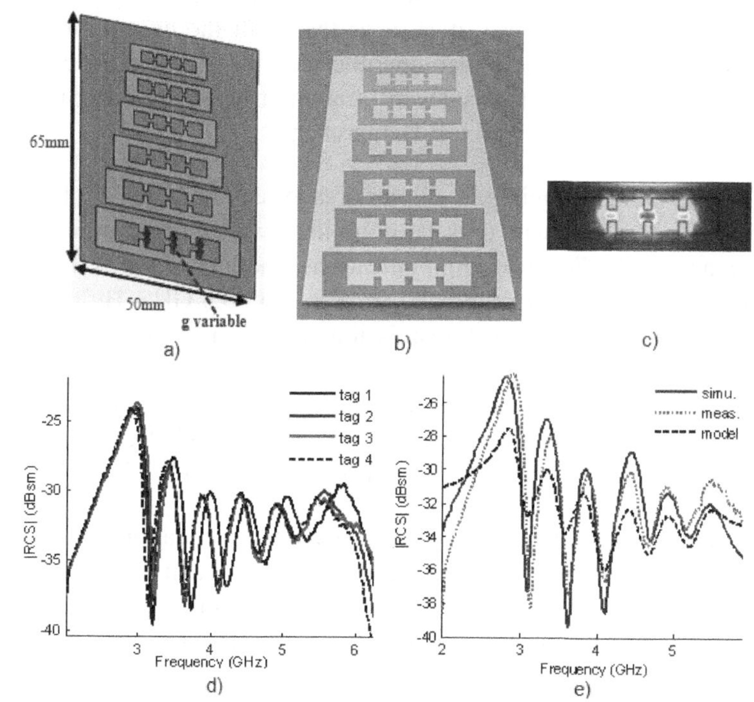

Figure 5.17. *Chipless tag designed to be printed via flexography on paper and corresponding measurement results. a) Schema and dimensions of tag designed; b) photograph of tag printed on glazed paper ($\varepsilon_r = 3$, tan δ = 0.095 at 2.5 GHz, thickness of 220 μm), c) visualization of the concentration of the electric field at resonance (the red zones are areas where the field is larger); d) comparison of RCS measurements for 4 different configurations obtained by causing the gap g to vary; e) simulation–measurement comparisons with an analytical model developed for the design phase. For a color version of this figure, see www.iste.co.uk/perret/ radio.zip*

Whether we are considering temporal or frequential chipless tags, all of these tags can be read using a temporal reading system [RAM 11, VEN 11b, VEN 11d] due to the time–frequency duality. The principal interest of the use of a temporal approach has to do with compliance with the rules regulating the use of frequency bands [VEN 11a]. As we have seen, chipless tags – particularly frequential ones – consume a great deal in terms of frequency band. The quantity of information they can contain is directly related to the frequency band that can be associated with them. If we look at the frequency bands that are free to use, it is clear that with the exception of the UWB approach, these are incompatible with the use of most chipless tags. For this reason, it is relevant to attempt to ensure compatibility between chipless RFID and UWB approaches.

Based on this principle, a chipless reader can be seen as a form of UWB radar, which will thus be compliant with current regulations. For example, the Federal Communications Commission (FCC) defines a SPD of -41.3 dBm for the band from 3.1 to 10.6 GHz. By reducing the cyclical ratio of pulse transmissions as much as possible, a very brief impulse (with a width of less than 100 ps) with elevated power (several volts on 50 ohms) can be sent by the reader. This technique has been experimented with for the reading of both frequential and temporal chipless tags [VEN 11a, VEN 11b, VEN 11d, VEN 12b].

The main difficulty in enabling a practical application here has to do with the reception stage of the chipless reader. In order to obtain the backscattered signal, the sampling frequency must be much higher than the highest frequency contained in the received signal [AGI 14]. For UWB signals, this sampling frequency must be around 40 GS/s. The operating principle of this type of reader is shown in Figure 5.18. However, performances like this are currently accessible only with laboratory devices such as wide-band oscilloscopes, which are much too costly to be reused in a demonstrator. It has become necessary to find alternative solutions. One promising solution is based on equivalent-time sampling, the principle of which is illustrated in Figure 5.19. Unlike classic communications systems based on temporal coding techniques, chipless tags are stationary devices. Next, we can attempt to reconstitute the whole signature of a tag by taking samples of different pulses sent back, rather than a single one. A reader based on an architecture using equivalent-time sampling is therefore a viable possibility. The main advantage here is that there is no need for a rapid analog-to-digital converter (see Figure 5.19). This would be a lower-cost solution, as it is sufficient to use delay generators producing delays on the order of a dozen ps. In this case, the sampling frequency can be low (a few dozen MHz). The sampler must, however, be capable of blocking the sampled value for a time without deforming it (meaning a wide frequency band). A reader of this type has been realized and tested in a real usage environment; Figure 5.20 shows the acquisition of a UWB pulse produced with a 12 GHz oscilloscope and the reader developed (equivalent time approach, Figure 5.19). We can see that there is a very good comparison between the two temporal signals.

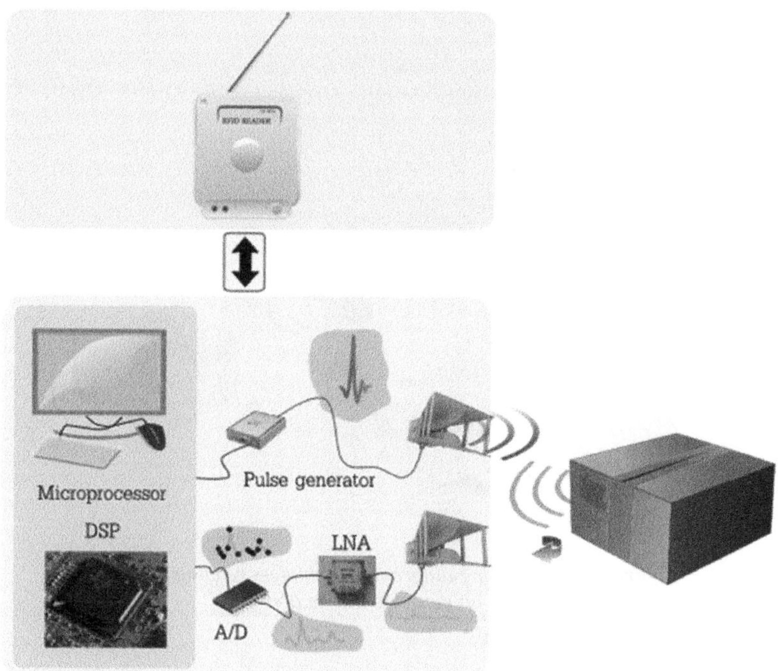

Figure 5.18. *Operating diagram of a temporal chipless reading system*

Other all-inclusive approaches are also possible here. Devices have recently become available for localization applications, notably through obstacles. These devices are simply UWB radars, some of which can be converted into chipless readers. For example, the UWB radar from Novelda AS has been used to read chipless tags (see Figure 5.20) [GIR 12, VEN 13b]. The operation of this radar is based on a different concept from equivalent time; it uses a paradigm called Continuous Time Binary Valued [RAD 13].The latter is asynchronous, and the return channel possesses a battery of 1-bit analog–digital converters. On the electronic board, a microcontroller is used to pilot the front-end RF and to link with a PC on which a localization application can operate.

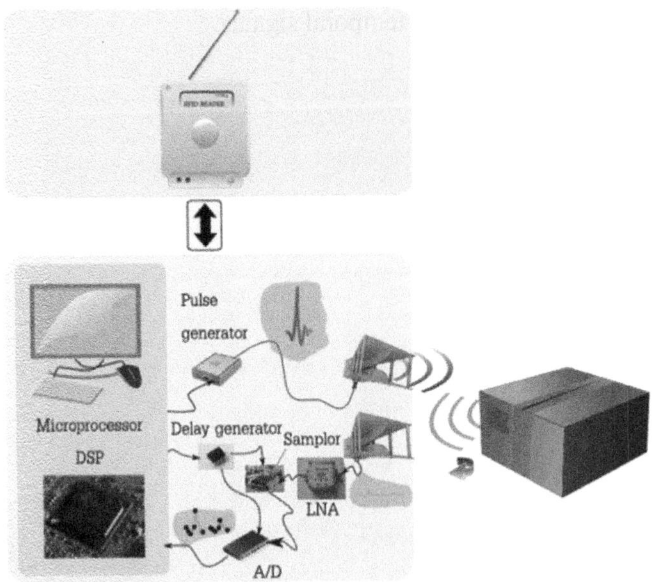

Figure 5.19. *Operating diagram of a chipless reading system based on an equivalent-time approach*

Thus radar makes it possible to achieve sample rate of around 35 GS/s at an accessible cost and compliant with European or FCC regulations. For all of these approaches, an averaging operation carried out on a hundred measurements is used to increase the signal-to-noise ratio of the radar. Figure 5.21 shows a comparison between measurements taken with a VNA and with Novelda radar. Here again, we see the characteristic peaks coding information, which validates the use of this type of device to produce a chipless demonstrator.

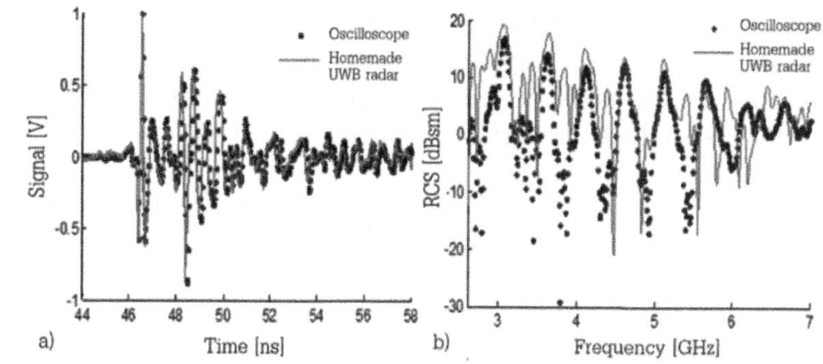

Figure 5.20. *Comparison between measurements taken with a portable low-cost reader developed on the equivalent-time principle and the acquisition yielded by the 12 GHz oscilloscope operating in real time. a) Temporal signals reflected by the tag shown in Figure 5.4(b), b) RCS obtained without calibration, solely by subtracting the signal from the empty chamber*

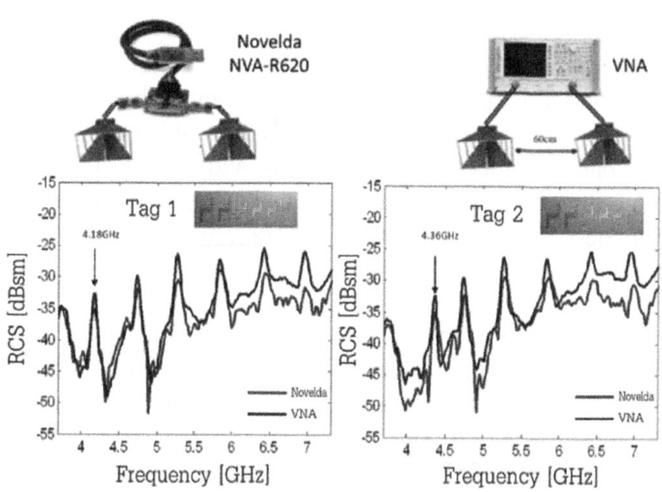

Figure 5.21. *Comparison between the reference measurement made with a VNA and the one obtained with the Novelda reader for two configurations of the same tag (only the first peak is moved, as expected). For a color version of this figure, see www.iste.co.uk/perret/radio.zip*

5.4.4. *Chipless RFID at THz – the THID project*

To end this chapter we will mention the possibility of producing chipless tags at higher frequencies, such as Terahertz (THz). After a brief introduction to the principal motivations for this, we will examine the results obtained at THz frequencies, and then introduce the first chipless tags composed exclusively of dielectric materials, used as part of the THID project, developed by Fréderic Garet's team at the IMEP-LAHC laboratory and realized by the CTP.

We have discussed the problem of RFID tag cost at some length. A second obstacle to the development of RFID (at least on a smaller scale) concerns challenges around the securing of the data present in the tags. Currently, classic RFID chips, the cost of which is minimal, lack reliability and security with regard to the information they contain. To resolve this double issue, we have begun with the observation that the perfection of low-cost tags that cannot be falsified is highly desirable today. It would be of particular interest, in fact, to be able to read an identification code with information on the content of any box – knowing that this code could not be read or modified by any unauthorized person – even if only from a short distance away. Here again, the production of chipless tags constitutes a very attractive solution. Thus we have worked on a new generation of chipless tags whose information is recorded in both surface and volume. The surface information is as we have described throughout this chapter; that is, a conductive pattern deposited on the surface and read by a UWB signal. The volume information has been produced using a multilayer structure whose thicknesses are compatible with THz frequencies; this part of the information is read by THz signals.

In terms of fixed specifications, the tag should have a surface area of a few square centimeters and a thickness on the order of one millimeter. The user will have three possibilities to memorize their information: surface only (RF), only in THz, or on surface and in volume at the same time (THID). Note here linking both types of information contributes a very interesting level of flexibility and additional coding, increasing the levels of both safety and security. In applicative terms, notably for reasons of confidentiality, it is expected to be able to code information directly in the material; that is, in the volume. These two approaches (surface and volume coding) are different from a technical point of view, but the information coding principle remains the same. The identification code of the tag is still based on the generation of a specific EM signature. The difference here is that, on the one hand, it is the conductive pattern that imposes a specific signature, while on the other hand, it is a case of stacking dielectric layers. Information coding can thus take place on the surface and in the volume as well. For this reason, the information is recorded directly in the substrate, which makes it unmodifiable and inalterable.

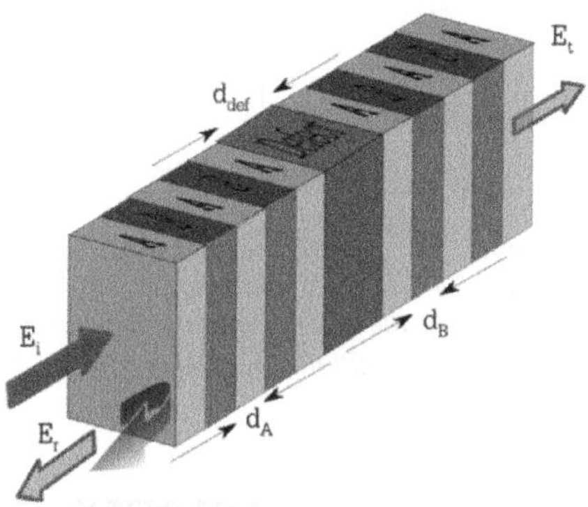

Figure 5.22. *Operating diagram of a THz chipless tag. The layer labeled "Defect" disturbs the regular alternation of layers A and B, composed by a high and low index material, respectively. Reading is done in transmission in this case. However, for practical reasons, it is possible to add a reflective plane in the center of the structure in order to have reading via reflections, which reduces the tag's thickness by two*

The principle of coding in volume is described precisely in articles [BER 11a, BER 11b, PER 11, TED 10]. It is based on Bragg gratings; that is, a multilayer structure of alternating materials with varying refractive index that makes it possible to obtain range of wavelengths where the EM signal is "forbidden" to propagate in the structure. The simplest configuration is the alternation of two layers with an elevated index contrast. From there, it is known that if we disturb the structure by adding a defect (variation of the thickness of a layer or modification of the index), a peak (or a ridge of peaks) may appear in the forbidden band gap [NEM 04] (see Figure 5.22). The idea has been to use these peaks to code information. This possibility of causing the position of one or more defects in the forbidden band gap

[BER 11a, PER 11] results in a large number of different configurations that can be differentiated without ambiguity on a frequency band as large as several hundred GHz (see Figure 5.23). It has been shown that it is possible to attain around 20 bits using 4 independent defects [HAM 12, HAM 13, HAM 14].

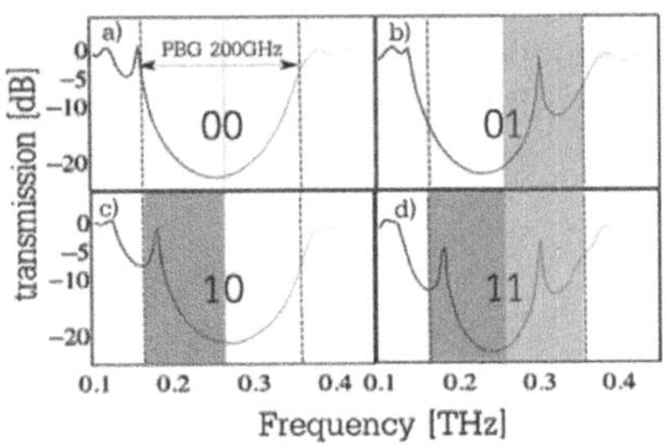

Figure 5.23. *Simulations of the transmission coefficient of four THz tags produced with stacks of silicon wafers (n_A=3.4, d_A= 75 μm) and of air (n_B=1, d_B=255 μm). The defects are obtained by changing the thickness of the central air layer, which makes it possible to obtain four different signatures. a) d_{def} = 230 μm, b) d_{def} = 460 μm, c) d_{def}= 125 μm, d) d_{def} = 955 μm*

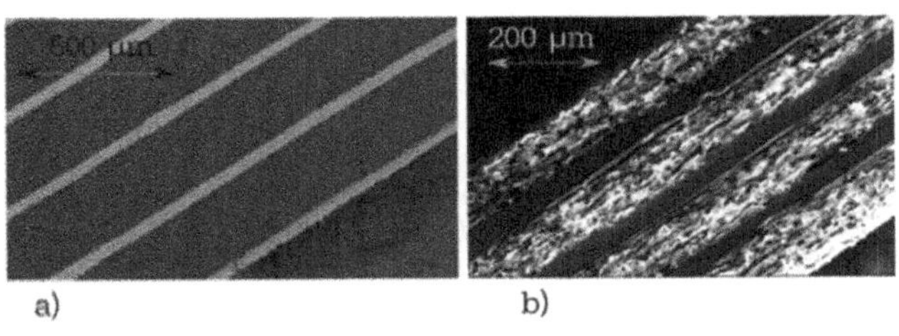

Figure 5.24. *SEM views of the stacks of layers used to produce a THz tag. a) PE-based tag, b) paper-based tag*

Figure 5.25. *Photograph of two THz tags: a) made of PE, b) made of paper*

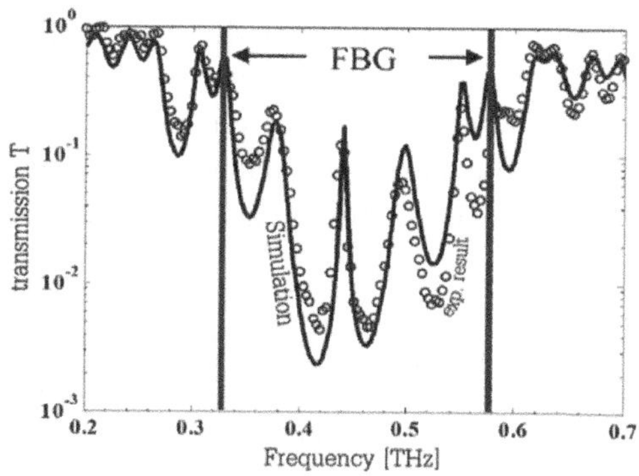

Figure 5.26. *Simulation and measurement of the transmission coefficient of the PE tag shown in Figure 5.25(a)*

The first structures were made by stacking silicon wafers separated by air gaps [BER 11a]. These were followed by structures produced using paper and polymer materials, as shown in Figures 5.24 and 5.25 [HAM 12, HAM 11, PER 13]. The high-index layers were produced by adding loads (TiO_2 for example) classically used in the paper industry. The results of measurements of a tag produced using polyethylene (PE) are shown in Figure 5.26. Figure 5.27 shows a tag including double coding; that is, both on the surface (RF approach) and in volume (THz approach).

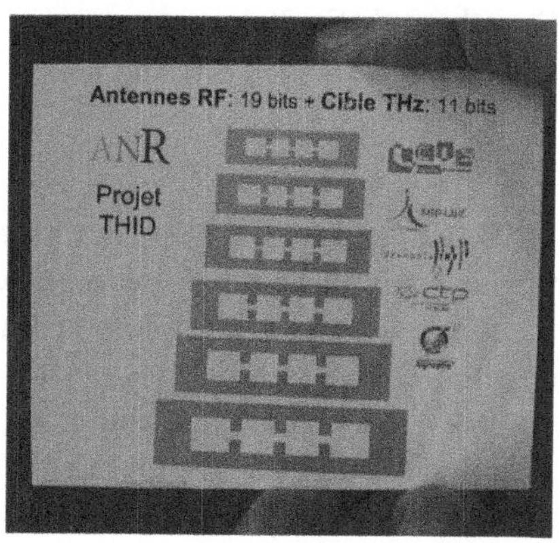

Figure 5.27. *Photograph of a THID tag, including surface information (RF approach) and volume information (THz approach)*

In summary, we can conclude that the principal motivations for this study in the THz domain are (1) increased security of the data contained in the tag; (2) the low-cost character of the solution (no conductive ink used; procedures and materials that are standard in the paper industry); (3) increased coding capacity and (4) new flexibility of use.

Concerning the first point, it is clear that any reverse engineering approach is difficult to apply, as it is necessarily destructive. Moreover, the absence of metal in the structure coding the information limits recognition of the source of the information being coded. It is not enough to attempt to discover the geometric dimensions of a pattern using imaging techniques; in this case, we must also be able to look at the index of layers. Additionally, the use of a new technology like this one, which is accessible to a limited number of actors and has a weak range (absorption by rays of water vapor, low-power sources, etc.) is necessarily difficult to adopt.

With regard to the fourth point, flexibility of reading, we may say that the risk of information alteration having to do with merchandise handling or transport is limited. Likewise, the tag can be the very packaging of the object being tagged, which will facilitate its being read. To conclude, we can also note that this coding system can be associated with optic identifiers including barcodes, classic RFID, and even chipless RFID without difficulty.

5.5. Conclusion

Based on the REP technology we have introduced, a significant amount of work has involved the development of original technological approaches, with the goal of demonstrating the practical and economic potential of a certain vision of chipless RFID. Numerous obstacles have been removed and now, thanks to these advances, extremely low-cost chipless technology is no longer simply a concept, but a reality. Adapted solutions have been contributed to the main problems of sensitivity of detection. Likewise, coding density [VEN 11a], robustness of detection [VEN 12b, VEN 12c], reading orientation [VEN 12d], tag production cost [PER 13,VEN 13c,

VEN 13d], reader production cost [ALE 12], and the issue of compliance with transmission norms [VEN 11d], are all themes with high practical impact that have been addressed and for which a response has been provided. It has been shown for the first time that it is possible to produce chipless tags on a large scale and for a unit cost on the order of 0.4 euro cents; that is, in compliance with the projections through 2019 that we noted in Chapter 1 for chipless RFID tag prices. These tags are produced via flexographic printing on paper. The only difference between them and barcodes is the use of conductive ink that produces the specific EM signature of the tags.

A new type of chipless tag technology combining surface and volume coding has also been introduced. It should be noted that volume coding does not require any conductive material, which is a first. Moreover, it enables the creation of an authentication field, which remains a very interesting field of application for chipless RFID [HAM 14]. The THz domain is in full boom, and portable readers have recently appeared [ZOM 14]. It is reasonable to think that within the next few years, these commercial solutions will make it possible to produce a chipless THz reader at a price compatible with the applications targeted.

Alongside these projects, another axis of research has progressively unfolded; it has to do with adding other functionalities to identification while remaining compliant with this vision of chipless, which must remain printable. The first of these concerns the sensor aspect. We will see in the next chapter how it is possible to add this function to some of the tags we have introduced here. It happens that in comparison to what was discussed in Chapter 3, the absence of the RFID chip can have a very positive impact in terms of sensitivity. The second sought-after functionality is the possibility of rewriting the tag's identifier. This service is also particularly desirable in the field of identification; even more so because it is impossible to implement it with barcodes, which clearly distinguishes this latest technology. We will move on, then, to an approach that is still highly prospective, which postulates an information rewriting function.

5.6. Bibliography

[AGI 14] AGILENT APPLICATION NOTE 1420 5988-8008EN, http://cp.literature.agilent.com/litweb/pdf/5988-8008EN.pdf, 2014.

[ALE 12] ALENCAR R., PERRET E., VENA A., "Construção de um Leitor para etiquetas RFID sem Chip", *15 SBMO – Simpósio Brasileiro de Micro-ondas e Optoeletrônica (MOMAG)*, 2012.

[AND 14] ANDRIAMIHARIVOLAMENA T., VENA V., PERRET E., *et al.*, "Chipless identification applied to human body", *IEEE International Conference on RFID-Technology and Applications*, Tampere, Finland, pp. 241–245, 2014.

[BAL 09] BALBIN I., KARMAKAR N.C., "Phase-encoded chipless RFID transponder for large-scale low-cost applications", *IEEE Microwave and Wireless Components Letters*, vol. 19, no. 8, pp. 509–511, 2009.

[BER 11a] BERNIER M., GARET F., PERRET E., *et al.*, "Terahertz encoding approach for securedchipless radio frequency identification", *Applied Optics*, vol. 50, no. 23, pp. 4648–4655, 2011.

[BER 11b] BERNIER M., PERRET E., GARET F., *et al.*, "Nouveaux tags RFID sans puce fonctionnant dans le domaine Terahertz", *17èmes Journées Nationales Micro-Ondes*, Brest, France, 2011.

[BLI 09] BLISCHAK A., MANTEGHI M., "Pole residue techniques for chipless RFID detection", *Antennas and Propagation Society International Symposium, (APSURSI'09), IEEE*, pp. 1–4, 2009.

[CHA 06] CHAMARTI A., VARAHRAMYAN K., "Transmission delay line based ID generation circuit for RFID applications", *IEEE Microwave and Wireless Components Letters*, vol. 16, no. 11, pp. 588–590, 2006.

[ETS 08] ETSI, "TS 102 754 V1.2.1 – Electromagnetic compatibility and radio spectrum matters (ERM); short range devices (SRD); technical characteristics of detect-and-avoid (DAA) mitigation techniques for SRD equipment using ultra wideband (UWB) technology", 2008 (available at http://www.etsi.org/deliver/etsi_ts/102700_102799/102754/01.02.01_60/ts_102754v010201p.pdf).

[GIR 12] GIRBAU D., LÁZARO A., RAMOS Á., "Time-coded chipless RFID tags: design, characterization and application", *IEEE International Conference on RFID-Technologies and Applications (RFID-TA)*, Nice, France, pp. 12–17, 2012.

[GUP 10] GUPTA S., NIKFAL B., CALOZ C., "RFID system based on pulse-position modulation using group delay engineered microwave C-sections", *Asia-Pacific Microwave Conference (APMC)*, Yokohama, Japan, pp. 203–206, 2010.

[HAM 11] HAMDI M., GARET F., DUVILLARET L., *et al.*, "THID tags for identification in the THz domain", *6èmes Journées Térahertz,* La Grande Motte, France, 2011.

[HAM 12] HAMDI M., GARET F., DUVILLARET L., *et al.*, "New approach for chipless and low cost identification tag in the THz frequency domain", *IEEE International Conference on RFID-Technologies and Applications (RFID-TA)*, Nice, France, pp. 24–28, 2012.

[HAM 13] HAMDI M., GARET F., DUVILLARET L., *et al.*, "Démonstration d'un Tag Chipless bas coût pour l'identification dans le domaine THz", *18èmes Journées Nationales Microondes*, Paris, France, 2013.

[HAM 14] HAMDI M., "Conception de tags d'identification sans puce dans le domaine THz", PhD thesis, University of Grenoble, 2013.

[HAR 07] HARTMANN C., CLAIBORNE L., "Fundamental limitations on reading range of passive IC-based RFID and SAW-based RFID", *IEEE International Conference on RFID*, pp. 41–48, 2007.

[HU 10] HU S., ZHOU Y., LAW C., *et al.*, "Study of a Uniplanar Monopole Antenna for Passive Chipless UWB-RFID Localization System", *IEEE Transactions on Antennas and Propagation*, vol. 58, pp. 271–278, 2010.

[IDT 14] IDTECHEX, http://www.idtechex.com/, 2014.

[JAL 05a] JALALY I., ROBERTSON I., "Capacitively-tuned split microstrip resonators for RFID barcodes", *35th European Microwave Conference,* Paris, France, pp. 4–6, 2005.

[JAL 05b] JALALY I., ROBERTSON I., "RF barcodes using multiple frequency bands", *IEEE MTT-S International Microwave Symposium Digest*, 2005.

[MAN 09] MANDEL C., SCHUSSLER M., MAASCH M., *et al.*, "A novel passive phase modulator based on LH delay lines for chipless microwave RFID applications", *IEEE MTT-S International Microwave Workshop on Wireless Sensing, Local Positioning, and RFID IMWS*, Croatia, pp. 1–4, 2009.

[MCV 06] MCVAY J., HOORFAR A., ENGHETA N., "Theory and experiments on Peano and Hilbert curve RFID tags", *Defense and Security Symposium*, pp. 624808–624808-10, 2006.

[MUK 07] MUKHERJEE S., "Chipless radio frequency identification by remote measurement of complex impedance", *European Microwave Conference*, pp. 1007–1010, 2007.

[NAI 12] NAIR R., PERRET E., TEDJINI S., "Temporal multi-frequency encoding technique for chipless RFID applications", *IEEE MTT-S International Microwave Symposium Digest (MTT)*, Montreal, Canada, pp. 1–3, 2012.

[NAI 13a] NAIR R.S., "Contribution to the development of time domain chipless tags and sensors", PhD thesis, University of Grenoble, 2013.

[NAI 13b] NAIR R.S., PERRET E., TEDJINI S., "A temporal multi-frequency encoding technique for chipless RFID based on C-Sections", *Progress in Electromagnetics Research B*, vol. 49, pp. 107–127, 2013.

[NAI 13c] NAIR R.S., PERRET E., TEDJINI S., *et al.*, "A group delay based chipless RFID humidity tag sensor using silicon nanowires", *IEEE Antennas and Wireless Propagation Letters*, vol. 12, pp. 729–732, 2013.

[NEM 04] NEMEC H., DUVILLARET L., QUEMENEUR F., *et al.*, "Defects modes caused by twinning in one dimensional photonic crystal", *Journal of the Optical Society of America B*, vol. 21, pp. 548–553, 2004.

[NIJ 12] NIJAS C.M., DINESH R., DEEPAK U., *et al.*, "Chipless RFID Tag Using Multiple Microstrip Open Stub Resonators", *IEEE Transactions on Antennas and Propagation*, vol. 60, pp. 4429–4432, 2012.

[NOV 13] NOVELDA, "Novelda nanoscale impulse radar", 2013 (available at http://www.novelda.no/).

[PER 10] PERRET E., TEDJINI S., DEEPU V., *et al.*, "Etiquette RFID passive sans puce", Brevet d'invention, FR 2956232 (B1), WO 2011098719 (A2), 2010.

[PER 11] PERRET E., HAMDI M., VENA A., *et al.*, "RF and THz identification using a new generation of chipless RFID tags", *Radioengineering-Special Issue: Emerging Materials, Methods, and Technologies in Antenna & Propagation*, vol. 20, no. 2, pp. 380–386, June 2011.

[PER 13] PERRET E., HAMDI M., TOURTOLLET G.E.P., *et al.*, "THID, the next step of chipless RFID", *IEEE International Conference on RFID*, Florida, pp. 261–268, 2013.

[PRE 08] PRERADOVIC S., BALBIN I., KARMAKAR N., *et al.*, "Chipless frequency signature based RFID transponders", *European Conference on Wireless Technology (EuWiT'08)*, pp. 302–305, 27–28 October 2008.

[PRE 09] PRERADOVIC S., KARMAKAR N., "Design of fully printable planar chipless RFID transponder with 35-bit data capacity", *European Microwave Conference*, pp. 13–16, 2009.

[RAD 13] NOVELDA NANOSCALE IMPULSE RADAR, http:// www.novelda.no/, 2013.

[RAM 11] RAMOS A., LAZARO A., GIRBAU D., *et al.*, "Time domain measurement of time-coded UWB chipless RFID tags", *Progress in Electromagnetic Research*, vol. 116, pp. 313–331, 2011.

[SHR 09] SHRESTHA S., BALACHANDRAN M., AGARWAL M., *et al.*, "A chipless RFIDsensor system for cyber centric monitoring applications", *IEEE Transactions on Microwave Theory and Techniques*, vol. 57, no. 5, pp. 1303–1309, 2009.

[TED 10] TEDJINI S., PERRET E., DEEPU V., *et al.*, "Chipless tags for RF and THz identification", *4th European Conference on Antennas and Propagation (EuCAP'10)*, Barcelona, Spain, pp. 1–5, 2010.

[TED 12] TEDJINI S., PERRET E., VENA A., *et al.*, "Chipless and Conventional Radio Frequency Identification: Systems for Ubiquitous Tagging", in NEMAI CHANDRA Karmakar (ed.), *Chapter 8: Mastering the Electromagnetic Signature of Chipless RFID Tags*, pp. 146–174, 2012.

[VEN 11a] VENA A., PERRET E., TEDJINI S., "Chipless RFID tag using hybrid coding technique", *IEEE Transactions on Microwave Theory and Techniques*, vol. 59, no. 12, pp. 3356–3364, 2011.

[VEN 11b] VENA A., PERRET E., TEDJINI S., "Novel compact RFID chipless tag", *Progress in Electromagnetics Research Symposium (PIERS'11)*, Marrakesh, Morocco, pp. 1062–1066, 2011.

[VEN 11c] VENA A., PERRET E., TEDJINI S., "RFID chipless tag based on multiple phase shifters", *IEEE MTT-S International Microwave Symposium Digest (MTT)*, Baltimore, MD, pp. 1–4, 2011.

[VEN 11d] VENA A., SINGH T., TEDJINI S., *et al.*, "Metallic letter identification based on radar approach", *General Assembly and Scientific Symposium, 2011 XXXth URSI*, Istanbul, pp. 1–4, 2011.

[VEN 12a] VENA A., "Contribution au développement de la technologie RFID sans puce à haute capacité de codage", PhD thesis, University of Grenoble, Valence, France, 2012.

[VEN 12b] VENA A., PERRET E., TEDJINI S., "Design of compact and auto compensated single layer chipless RFID tag", *IEEE Transactions on Microwave Theory and Techniques*, vol. 60, no. 9, pp. 2913–2924, 2012.

[VEN 12c] VENA A., PERRET E., TEDJINI S., "A fully printable chipless RFID tag with detuning correction technique", *IEEE Microwave and Wireless Components Letters*, vol. 22, no. 4, pp. 209–211, 2012.

[VEN 12d] VENA A., PERRET E., TEDJINI S., "High capacity chipless RFID tag insensitive to the polarization", *IEEE Transactions on Antennas and Propagation*, vol. 60, no. 10, pp. 4509–4515, 2012.

[VEN 12e] VENA A., PERRET E., TEDJINI S., "A compact chipless RFID tag using polarization diversity for encoding and sensing", *IEEE International Conference on RFID*, Florida, pp. 191–197, 2012.

[VEN 13a] VENA A., PERRET E., TEDJINI S., "Design rules for chipless RFID tags based on multiple scatterers", *Annals of Telecommunications,* Special Issue on Chipless RFID, vol. 68, nos. 7–8, pp. 361–374, 2013.

[VEN 13b] VENA A., PERRET E., TEDJINI S., "A depolarizing chipless RFID tag for robust detection and its FCC compliant UWB reading system", *IEEE Transactions on Microwave Theory and Techniques,* vol. 61, no. 8, pp. 2982–2994, August 2013.

[VEN 13c] VENA A., PERRET E., TEDJINI S., *et al.*, "Conception de tags RFID sans puce imprimés sur papier par Flexographie", *18èmes Journées Nationales Microondes*, Paris, France, 2013.

[VEN 13d] VENA A., PERRET E., TEDJINI S., *et al.*, "Design of chipless RFID tags printed on paper by flexography", *IEEE Transactions on Antennas and Propagation,* vol. 61, no. 12, pp. 5868–5877, 2013.

[ZHA 06] ZHANG L., RODRIGUEZ S., TENHUNEN H., *et al.*, "An innovative fully printable RFID technology based on high speed time-domain reflections", *High Density Microsystem Design and Packaging and Component Failure Analysis (HDPapos'06)*, Shanghai, pp. 166–170, 27–28 June 2006.

[ZOM 14] ZOMEGA, http://www.zomega-terahertz.com/, 2014.

6

Perspectives on Chipless RFID Technology

This final chapter presents ongoing developments, as well as perspectives on chipless radio frequency identification (RFID) having to do with the idea of adding different functionalities to the chipless RFID tags previously introduced.

6.1. Introduction

The need for information identification and capture is a subject of utmost importance in modern societies. Every sector of society relies on the identification and safeguarding of data exchanged, the updating of data recorded on a tag and the measurement of physical parameters such as humidity or temperature. This is especially true in the field of traceability of commercial items and in the medical industry. The ability to make objects interact with one another and/or with humans is an important factor in many applications, all the more so if this interaction can occur without human presence.

In this book, we have examined two approaches that are used to transmit ID remotely, both based on an RF link. Chapter 3, which focused on the possibility of developing functions to sense physical values from classic RFID tags alone, showed that it is possible to graft other features onto this type of already-extant communication system.

This principle based on *analog reading* can be transposed to chipless technology as well; thus, without adding specific components, it is possible to use traditional ultra high-frequency (UHF) RFID for sensor applications [CAI 11a, CAI 11b, MAR 11]. However, we can see that the difficult-to-reproduce behavior of the RFID chip for these types of applications (it was not originally made for them) will considerably limit the measurement of physical parameters [GUI 12]. Based on this, a solution that does not require a chip would make it possible to obtain much better performances in terms of sensitivity. This is only possible if the tag used to take the measurement also contains its own identifier. We can then, as in classic RFID, put a network of wireless – and, in this case, chipless – sensors in place. The development of chipless technology makes it possible to resolve this issue neatly. This is why an analogous approach, but applied this time to chipless RFID, has been adopted in the production of low-cost tag-sensors.

This chapter describes four developments toward which chipless technology is oriented. The first development concerns the issue of the securing of information that will, in chipless RFID, be able to be processed in the "physical layer". The second development describes the possibility of carrying out multiple readings. Then, the chipless RFID sensor issue will be introduced. The final section will be dedicated to the problem of rewriting information on a chipless tag. A potentially printable approach will be introduced, and perspectives on this new solution will be discussed.

6.2. Securing of information

Another very important aspect of wireless communication is the securing of data. Although things are changing, there is currently little (mainly in HF RFID) or even very little (as is the case in UHF RFID) security in RFID communication, which negatively affects its deployment in some areas. Despite all of this, chipless technology responds rather well to these expectations. Its information is non-modifiable, and more difficult to duplicate than the information in barcodes or classic RFID. The conductive deposit that codes information can easily be hidden by the addition of an opaque protective film, which enables discreet reading, which is difficult to implement with a barcode, for which it is necessary to handle the object, and whose code must be visible in order for its image to be captured. Discreet reading of an RFID tag is possible if the object is large enough to be able to hide the tag. It should be noted here that chipless RFID is highly non-intrusive, and it is relatively simple – as with barcodes – to find a surface (particularly within the packaging) on which to place the tag, with the major benefit here that it can be covered up and thus invisible. Likewise, chipless technology does not affect the integrity of the product being identified, unlike barcodes, which must remain visible. Classic RFID requires the presence of a tag containing a silicon chip, and most often a conductive part created on a plastic substrate, which can be problematic in terms of both product integrity and the recycling of the materials.

If we continue to list the differences between these approaches in terms of the security aspect, it also becomes clear that, unlike RFID (or surface acoustic wave (SAW) technology), in which a strong field can destroy the chip (or the piezoelectric material), the chipless approach is resistant to destructive attacks of this kind. Another point is the reading zone, which remains much more limited than in UHF RFID. It is actually easier to delineate, which may make it possible to strengthen the level of communication significantly, or simply to ensure the acquisition of the ID of the tag (or several tags) present in the reading zone, thus avoiding the problems of cross-reading that are quite common in RFID. The act of reading a tag other than the one desired is very harmful, which is why more and more research is focusing on the development of methods to impose this reading zone [NIK 07, DAI 14]. Systems in which reading must be guaranteed in a well-delineated zone are highly desirable today, and chipless technology may provide a solution. We are seeking then to guarantee an optimal read rate in this zone. Conversely, any tag that does not fall within this zone must not be read. HF RFID is currently providing a solution to this problem; however, in this case, the reading zone is small (a read range of around a dozen centimeters); the dimensions of the tags are large, and the read rate is lower than in UHF. Moreover, the number of simultaneous readings is reduced and the production cost of the tags is markedly higher than in UHF. For this reason, it is of importance to imagine the use of passive UHF RFID for these applications in the near field. With regard to chipless technology, we may seek to obtain a reading method similar to that of HF RFID, but with lower prices and better miniaturization of tags.

Still with regard to security, another point has to do with the mode of interaction chosen, that is the use of an open- or closed loop. In the second case, we may imagine keeping the coding part as private. The recovery of the ID is linked to the decoding stage, which enables access to the information contained by the tag (link between the position of the determining elements of the signal and the corresponding binary code). If the method of coding information is public, as with barcode and RFID technologies, any person with a reader who is near the tag is able to read the information. If the code is not made public, reading is secured.

To conclude, elements linked to the security specifically of multilayer tags functioning in the THz domain were discussed in the last chapter. We have seen that volume coding without conductive elements is of interest from this point of view. In this case, any reverse-engineering approach is more complex to implement, which will go a long way toward limiting the counterfeiting of tags.

6.3. Multiple readings

Volumetric reading, that is the reading of a set of tags, simultaneously and in a given space, is a sought-after functionality in a large number of applications [NIK 07]. Barcodes do not provide a solution to this issue; only UHF RFID is able to do so in this case, where several hundred to a thousand tags can be read at virtually the same time. However, as previously mentioned, it is difficult to set the parameters of the reading zone without using a Faraday cage, especially when metallic objects are present. The chipless approach may contribute a modest solution to the issue of multiple readings. It should be noted that this functionality is not incompatible with chipless technology, which sets it apart from barcodes. It is still possible to read several chipless tags at the same time, but it must be said that this is at the cost of a significant reduction in coding capacity. To do it, it is necessary to force all of the signatures to be orthogonal to one another; a simple way to do this consists of attempting to produce signatures defined on intervals that are separated from one another, thus making it possible to receive multiple signals backscattered by multiple tags simultaneously without losing the identifier sent back by each of these tags [PER 14].

In temporal systems, the idea is to play on different delays for each tag. Thus, a group of tags located in the same reading zone can be designed to send their specific signature over different temporal slots; this is the technique used in SAW technology. The difficulty is to be able to generate significant delays linked to the zone being covered and the number of tags we wish to read at the same time. The delay used here to prevent any collision cannot be used to code information, which will necessarily limit the number of bits we can code. Note, however, that for relatively small reading zones (for example a zone of around 30 cm on each side), the difference in time-of-flight is smaller than one nanosecond, which is the same value as the temporal range classically associated with each chipless tag (see temporal tags based on the C-sections described in Chapter 5 [NAI 13a]). Thus, we can see that by limiting the reading zone, we can read several tags of this type without reducing the information quantity too much [PER 14].

In frequential systems, a simple way of solving the problem is to associate a frequential range of its own with each tag. In this case, a tag can be reduced to a single resonator that will have only one resonance frequency. If the tags are far enough apart from one another to avoid coupling, multiple tags will be fully detectable in the same zone. However, information quantity will be greatly reduced in this case, with the total number of configurations being neither more nor less than the number of frequential slots able to be used. Due to the 100 MHz of range per resonator, with around 7 GHz of total band (in ultra wide band (UWB)), we will have 70 different tags at most. However, it will be possible to read all of these tags simultaneously.

6.4. Chipless sensors

In applicative terms, the desired objective is the perfection of a new generation of sensors that are identifiable, easy to use, and potentially low-cost, able to fulfill the pressing need to make objects able to communicate with one another. This recently appeared theme is known as the Internet of Things (IoTs). This is above all an effort to extend the Internet into the sphere of places and things, using tags to ensure this link. The ability to grant information access (identification and sensor functions) via an RF link, via objects from the daily life of a reader who is himself or herself connected to the Internet, would enable a large number of new, previously unknown applications. Thus, the perfection of low-cost tag-sensors is highly sought after today. It would be of particular interest to be able to read an identification code providing information about the content of an object – data on its hygrometry, for example. We would have access to an object-tracking system that would be remote and extremely complete, all using low-cost technology. For this, the production of chipless tag-sensors is a very attractive solution. Compared to the classic RFID solution introduced in Chapter 3, besides increased precision, the ability to avoid chip-related constraints reduces cost, increases lifespan and results in tags that are more mechanically robust overall, with much higher resistance to vibration and temperature. We will see how it is possible to envision the development of a new generation of chipless tag-sensors containing an identifier. Two types of sensors will be described here: the first would be used to measure humidity, while the second would be used to measure the geometric deformation of an object.

6.4.1. *Humidity sensors*

Based on chipless technology, the idea here is to add on a sensor function. Various materials can be used to do this, for example nanomaterials. Silicon nanowires have geometries and dimensions with a very high surface-to-volume ratio, thus encouraging surface interactions. Given the very small dimensions of these structures (the diameter of a nanowire can be on the order of a few dozen nanometers), exchanges or harnessing of molecules can take place on the surface, permitting a modification of electrical properties depending on the environment in which they are placed. These concepts have been studied for several years already, and the possibility of using them as sensors, and particularly wireless sensors, has been demonstrated.

6.4.1.1. *Latest developments*

Nanotechnologies are exciting a growing amount of interest in the field of sensors, since they allow the use of extremely sensitive materials [STA 06]. Recent work has demonstrated the effectiveness of carbon nanotubes in detecting nitrogen oxides [MCG 06], smoke and oxygen in very low concentrations. At the same time, conductive and semiconductive nanowires have been used for their detection capacities, which can vary depending on their stimulation or chemical composition. Semiconductive nanowires have widely been studied in the last decade as they display remarkable physical properties, notably a high surface-to-volume ratio, and the very high sensitivity of their intrinsic properties to physiochemical substances that may be adsorbed by their surfaces [CUI 01]. Several techniques are used to produce them, including advanced lithography, plasma engraving or growth from a metallic catalyst.

The physical properties of silicon nanowires show sensitivity to levels of humidity [LI 09a], temperature [CHU 11], nitrogen oxides [PEN 09] and hydrogen [SKU 10]. In every case, a variation of the conductivity or effective permittivity of the nanowires has been observed. More generally, the principle of the sensor function using a change in the conductivity of nanomaterials has been introduced and validated at low frequency [CHE 11, CHU 11, LI 09a, PEN 09, SKU 10]. However, most promising technology is the one that consists of detecting changes in the conductivity and permittivity of nanomaterials for radiofrequencies [LI 09b, MCG 06, YOO 05]. Proof on the theoretical principle of wireless sensors based on the use of nanomaterials has been given; [LI 09b] presents a wireless and chipless sensor using the principle introduced in [YOO 05]. We may note that carbon nanowires are deposited in an area corresponding to the input port of an antenna, which makes it possible to produce the wireless sensor. The presence of a gas causes a variance in the impedance seen by the antenna and, from there, any wave backscattered by the antenna contains the information in the presence of the gas. It should be noted, however, that no practical application (remote measurement of the variation in the physical parameter) is presented. On the contrary, the difficulties involved in taking measurements remotely (sensitivity of the RF channel in relation to the tag's environment, etc.) are emphasized [MCG 06]. Also, note that none of these papers envisions the association of an identifier in order to produce a tag-sensor. This point is of great importance from an applicative point of view, as it enables the creation of a network of sensors. It is a matter, then, of the extent to which working with RF frequencies has the undeniable benefit of making it possible to implement a (wireless) remote reading system. RFID technologies seem particularly adapted to the integration of sensitive nanomaterials. The fundamental advantage of RFID lies in the self-powering of tags and in wireless communication, rendering the exchange of information distinctly faster and more effective. Conversely, as we have seen, the presence of an RFID chip reduces the precision of measurements. The chipless RFID approach we have presented here is an elegant and effective solution, making it possible to provide, at lower cost, an identifier for the tag and an answer to all of the problems of wireless RF measurements. In terms of performance, in this case, the benchmark example is SAW technology, by which highly sensitive humidity sensors have been created [LIE 09, NOM 94, SHE 11]. To obtain this type of tag, the solution simply consists of adding to a classic SAW tag a material with hygroscopic properties. This material will modify the characteristics of the acoustic wave according to the rate of humidity present, modifications that can be picked up by the reader. However, the procedures used to produce SAW devices call on techniques developed in microtechnology, thus increasing both the complexity of manufacturing the sensor and, above all, the cost.

As in classic RFID, the question of adding a sensor function to chipless RFID has been raised, though more recently [JAT 07, AUB 09, JAT 10, PER 10b, THA 10]. In [BAL 08], we find a compact, wireless, chipless sensor used to detect ethylene. This sensor is based on the use of a condenser, which is the sensitive component and is directly integrated into the tag. The same team then tried to add an ID to the chipless tag using a delay line [SHR 09], but in this case (with the same thing true for [PER 10b]), the sensor function is simply reproduced by soldering an element located at the end of the delay line (several CMS capacitances of different values are used to simulate the variation in the physical value). One of the first totally integrable chipless tag-sensors, with remote (wireless) reading of the identifier and the value to be measured, was experimentally validated in 2011 [TED 11, VEN 12c]. By chipless tag-sensor, we mean a compact chipless tag with identification and sensor functions that are totally integrated and compatible with the spirit of chipless RFID (that is simple in design and low in cost, as well as potentially fully printable, similar to barcodes in that there is no discreet element connected to the tag). Silicon nanowires were used to obtain this result. They were initially in an aqueous solution, and subsequently deposited in drops on a classic chipless tag. This is the use of chipless technology, and especially of silicon nanowires, which is at the heart of this advance.

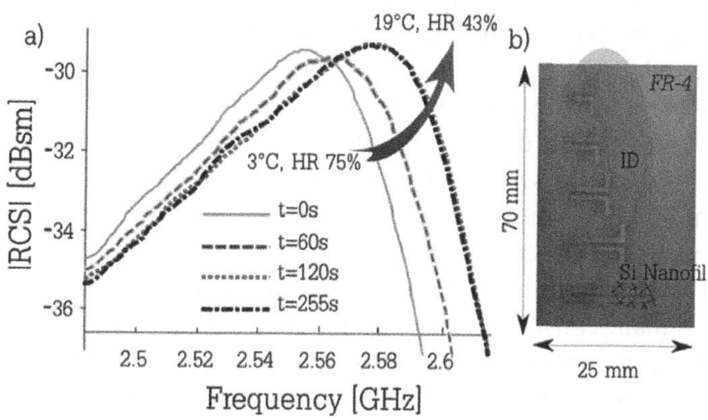

Figure 6.1. *a) Influence of humidity on the EM response of a resonator containing a deposit of Si nanowire. b) Photograph of a chipless RFID tag integrating the sensor function. For a color version of this figure, see www.iste.co.uk/perret/radio.zip*

6.4.1.2. *Operation of proposed tag-sensors*

The most widespread humidity sensors, in practice, are hygrometers with capacitive probes, whose dielectric properties vary according to the humidity in the air. The hygroscopic polymer film used to produce the capacitance adsorbs water molecules from the surrounding air until equilibrium is attained with the water vapor it contains. The measurement of this capacitance then gives the value of the relative humidity in the air. Classically, the change in capacitance is 0.2–0.5 pF for a variation of 1% of relative humidity in the air, and the capacitance has a value of between 100 and 500 pF for 50% humidity and a temperature of 25 °C [MET 14].

The tag-sensors presented here use the operating principle of the use of a material sensitive to humidity, with the idea of making it interact directly with the wave, which will question the tag and the backscattered part of which must also contain its identifier. We may wonder how to make these two completely different signals – that is the ID and the data pertaining to the capture of the physical value – work together. Several approaches are possible. We might use the polarization of waves, thus associating a value with each "channel". We might also dissociate them frequentially or temporally. The latter two approaches have been applied, using the C-shaped and C-section-based tags introduced in Chapter 5, respectively. For the frequential approach, the simplest solution consists of using one of the resonators to measure humidity (see Figure 6.1) [VEN 12b, VEN 12d]. Since each resonator has a different resonance frequency, we are certain that it will always be possible to separate the information pertaining to the ID from that pertaining to the physical value. Figure 6.1(a) shows the variation in resonance frequency of the largest resonator (see Figure 6.1(b)) according to humidity: the other resonances (not shown in the figure) remain constant, provided that the humidity-sensitive material has only been deposited on the element resonating at 2.54 GHz. In this approach, it is actually the properties of the sensitive material that will determine those of the sensor. It is sufficient for its physical parameters such as conductivity, permittivity or permeability to be sensitive to the environment in order to be able to collect information. For our chipless tags, a variation in conductivity undeniably causes a change in the response of the signal of a resonator in the tag, while a change in permeability or permittivity will result in a gap in its resonance frequency. To do this, silicon nanowires have been deposited in the slot of a resonator (Figure 6.1(b)). The deposit has been made manually; the nanowires are placed in an aqueous solution, and then, using a pipette, several drops have been placed on precise spots on the structure (see Figure 6.1(b)). As shown in Figure 6.1(a), measurements taken with a classic chipless test bench have revealed the sensitivity of the sensor to humidity. Measures are taken every 15 seconds to monitor the variation of the tag's radar cross-section (RCS) as a function of time. As we will see, measurements taken at fixed humidities and then fixed temperatures – with the other value being variable – have made it possible to isolate the source of the variations observed. We will now present the results obtained using the tag based on C-sections. We will see that the information related to the variation of the physical parameter will be present in the group delay (GD) taken from the electromagnetic (EM) signature of the tag, and thus we are using the hybrid frequential/temporal approach in this case.

The operating principle of the tag-sensor based on a delay line is shown in Figure 6.2. It resembles the one previously introduced in Chapter 5 for identification applications [NAI 11]. We will first introduce the operating principle by considering a classical delay line (Figure 6.2), which will then be generalized with a C-section configuration (Figure 6.3). As shown in Figure 6.2, the tag-sensor is composed of a transmission antenna, a delay line and an antenna that will retransmit the signal propagated along the line back toward the reader in cross-polarization. The delay generated is a direct function of the length of the line (labeled as l_n in Figure 6.2). By playing on this length, it is possible to temporally shift the antenna mode (the signal propagating in the line) in relation to the structure mode (quasi-optical reflection), and thus to code an identifier. By adding a specific sensitive material on the tag, the presence of humidity can cause a variation of the delay and therefore of the time separating the two signals previously described (see Figure 6.2). Thus, by linking each tag to a temporal window (that is a given length l_n), it is possible to code an identifier and measure the variation in humidity at the same time [NAI 11]. The same idea can be performed with better performances by using C-section as we can see in Figure 6.3

[PER 14]. The C-section plays the role of a miniaturized delay line, and is characterized by a GD that is a direct function of the length of the line (labeled as l_n in Figure 6.4) and of the frequency. As we have seen in Chapter 5, the dispersive behavior of this device allows us to increase its coding density. It has been shown that by adding silicon nanowires in a manner identical to what has already been presented, the presence of humidity causes a huge variation of the C-section's GD and therefore of the time separating the two signals previously described. Figure 6.4 shows the correspondence that exists between the frequential and temporal domains, where we can see how to read the information coming from the tag.

This is a clear example of how an RFID reader can interrogate multiple tags simultaneously. In this case, we need to differentiate each tag but also to retrieve the physical information of each, and, of course, without any communication protocol, as in classical RFID. As discussed previously (section 6.3), different anti-collision schemes in chipless RFID can be used. To separate the response of only one tag, we can adopt either the principle of discrimination in time or frequency [PER 14]. If we do it in time, the tag will induce a specific delay. Then, we can assign a time slot to each of the n tags. This is exactly what we have with the delay line's configuration (Figure 6.2). The same principle can be used in frequency: each of the n tags operating at a different frequency, the information on the presence of the tag can be obtained by identifying the peaks in the signal spectrum. Windows are assigned in frequency, and the peak belonging to one or the other of these bands determines its identifier. Note that in the example of hybrid time-/frequency-domain tag based on C-section, the frequency windowing is using wells to separate tags from one another. But humidity information is present in the variation of the GD for the specific frequency (F_1) attached to the tag (Figure 6.4(b)). It is also possible to capture this information by looking at the signal in time (after having performed filtering at frequency F_1). The humidity value, in this case, is given by the delay observed between the structural mode and the second observable peak (see Figure 6.4(c)).

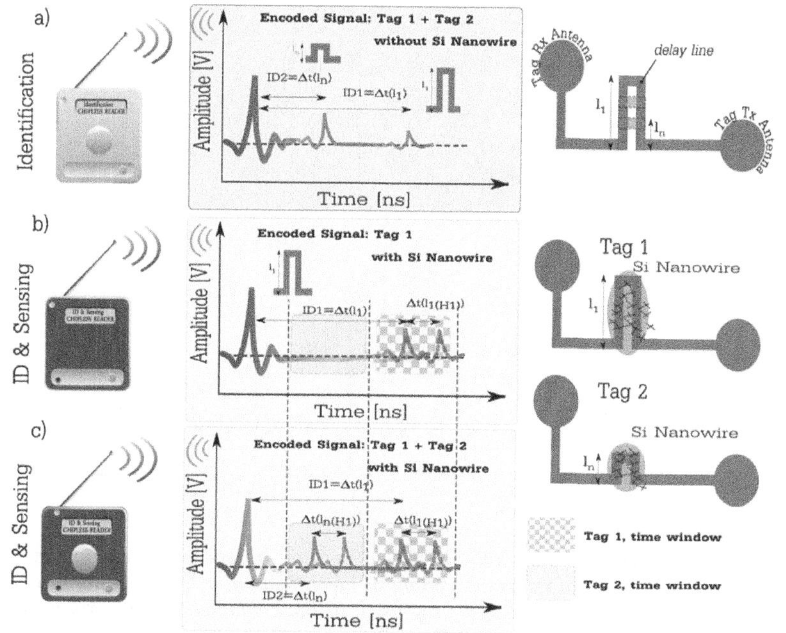

Figure 6.2. *Operating principle of a temporal tag-sensor. The chipless tag is based on the use of a variable delay line connected to two monopole antennas. a) Temporal coding of the ID. b) Recovery of the information measured (sensor function). c) Case of a network of tags; each tag is linked to a temporal window to enable the simultaneous reading of several tags. For a color version of this figure, see www.iste.co.uk/perret/radio.zip*

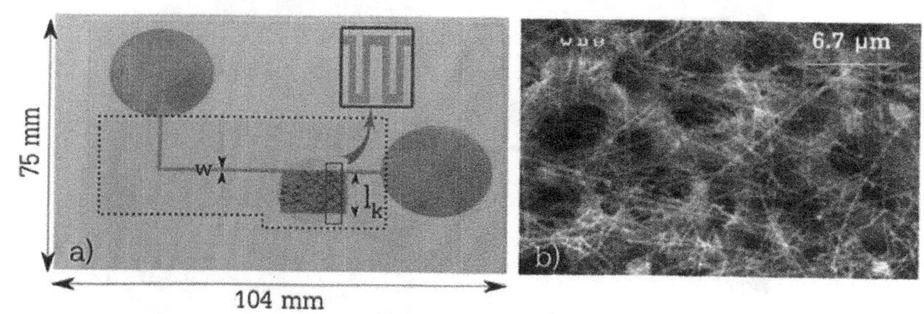

Figure 6.3. *Photograph of a tag-sensor realized using a C-section and nanowires. a) The chipless tag is composed of a group of 10 C-sections (w = 0.7 mm, l = 14.9 mm); the ground plane is indicated by the dotted line. The circle represents the area where the nanowires have been deposited. b) SEM image of silicon nanowires deposited between C-sections. For a color version of this figure, see www.iste.co.uk/perret/radio.zip*

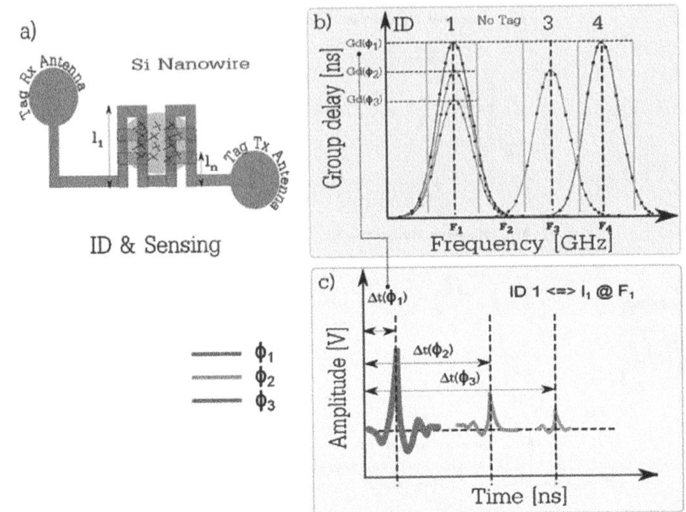

Figure 6.4. *Operating principle of the tag-sensor based on C-sections in the frequential and temporal domains: coding of identification and recovery of physical value φ. a) Operating schema of the tag used. b) Backscattered signals expressed in group delay versus frequency; three tags are read at the same time (1, 3 and 4). For tag 1, the group delay for three values of φ of humidity is represented. c) Backscattered signals, filtered at F_1 expressed in time for three values of φ. The information (ID and φ) is accessible both in frequency and time. For a color version of this figure, see http://www.iste.co.uk/perret/radio.zip*

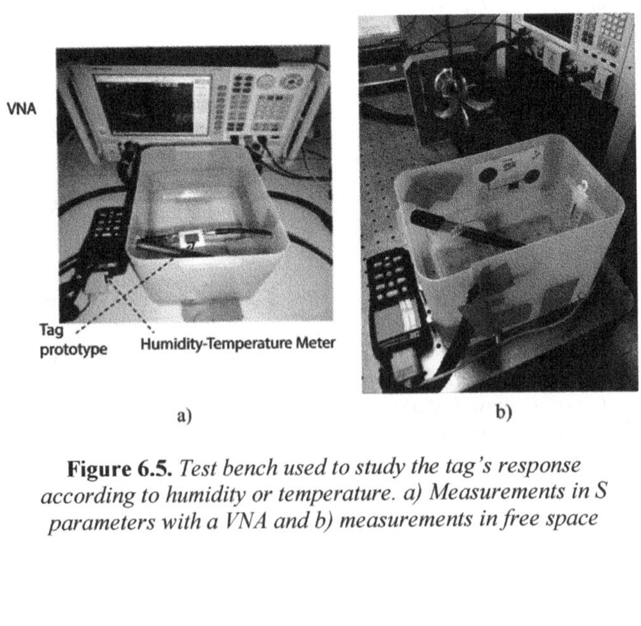

Figure 6.5. *Test bench used to study the tag's response according to humidity or temperature. a) Measurements in S parameters with a VNA and b) measurements in free space*

6.4.1.3. *Variation of humidity*

The results discussed here were obtained using the tag-sensor shown in Figure 6.3(a). They were produced at constant temperature, more precisely in a room climate controlled to 23 °C. A dozen drops of solutions containing silicon nanowires were deposited on the tag. The measurement bench is shown in Figure 6.5(b). A sealed container into which water was placed made it possible to cause the relative humidity (HR) to vary in time inside the box, between around 60% and 100%. The variations obtained in RCS and GD are shown in Figure 6.6. We can see significant variations around the characteristic frequency of the C-section ($l \approx \lambda/4$, or around 3.73 GHz): 40 MHz on the frequential position of the trough; 30 dB on RCS amplitude, and around 23 ns on the GD. If we attempt to write the relationship between the RCS and the relative humidity value, the curves shown in Figure 6.7 can be used to consider the evolution in time of these two values (for a frequency of 3.73 GHz). It is then possible to deduce from this the analytical relationship linking these two values. For this variation zone, the relative humidity is inversely proportional to the RCS (in decibels). Once these observations are made, it is a matter of interpreting the results. The difficulty is that we cannot attribute a thickness to the layer of nanowires present on the tag; this layer is extremely thin and difficult to characterize in practice; moreover, it is non-homogeneous, since the nanowires tend to attach themselves to the periphery of the deposited drops. Rather than trying to determine the permittivity of this deposit, for the simulations carried out, a simple model enabling access to the tendencies has been used. A box of thin dielectrics has been added to the simulation of the tag, covering the zone where the drops are deposited. The permittivity and the losses within this box have then been made to vary, in order to verify whether or not it is possible to recreate the behavior observed in practice. The results obtained are shown in Figure 6.8. We may note the variation of the tag's GD for different permittivity and loss tangent values. This gives us the behavior that is very similar to what is observed in practice: an increase in permittivity resulting in a translation toward the low frequencies of the GD peak, an increase in losses resulting in a reduced value of the peak until a signal change and then an increase of the latter (while remaining negative). We can see then that by combining the two variations (of permittivity and losses), it is possible to reproduce the tendencies present in the measurement results previously obtained. From this, we can deduce that the presence of humidity will simultaneously modify permittivity and losses in the area surrounding the C-section. A measurement of the same structure but without the presence of nanowires, under the same conditions, does not display any significant variation either in RCS or GD, which shows again that it is the nanowires that are exacerbating this phenomenon. A last important point is that it has been observed that the variations are reproducible from one day to the next. In fact, the same tag was measured under the same conditions several days later (or even after several weeks), and the results obtained are quite comparable, which means that the nanowires do not seem to move as time passes (either in terms of their physical properties or their placement on the tag).

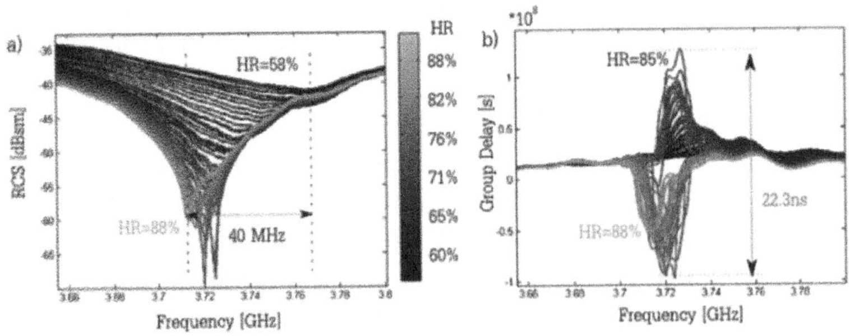

Figure 6.6. *a) Measurement of RCS and b) of group delay depending on the variation of the relative humidity between 60.2 and 88%, at ambient temperature for the chipless tag-sensor shown in Figure 6.3(a)*

6.4.2. *Deformation sensor*

We will now consider an example in which chipless RFID may contribute a solution to a concrete problem that we have been unable to address simply with other approaches until now [BEL 13]. The work presented here demonstrates that the measurement of the deformation, for example, of an antenna reflector positioned on a satellite, can potentially be done using chipless tags. After a brief description of the operating principle, we will move on to the solution chosen as well as the results obtained. An experimental validation will be presented; variations of less than 1 mm were measured.

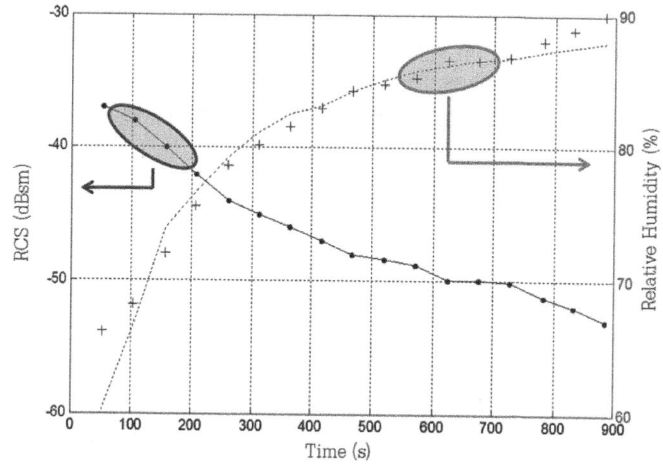

Figure 6.7. *Evolution in time of the measurement of the RCS of the tag-sensor and of the relative humidity. The dashed curve is obtained from the RCS curve by applying the relationship RH = a/RCS (dB) + b, where the coefficients a and b have been obtained via curve fitting (a = 2.918e + 3, b = 145)*

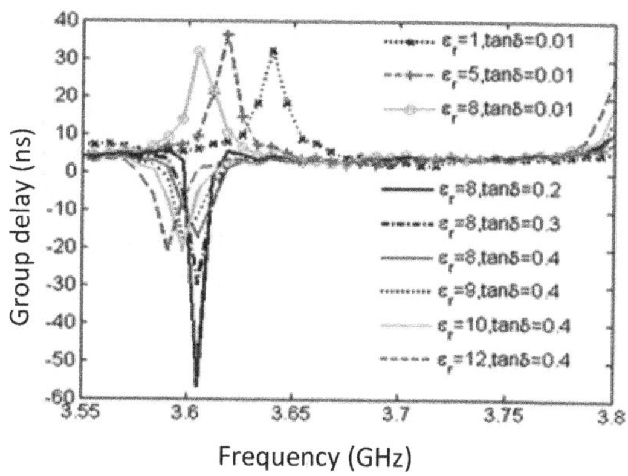

Figure 6.8. *Simulation of the variation in group delay of the chipless tag-sensor shown in Figure 6.3(a) for different permittivity (ε_r) and loss tangent ($\tan\delta$) values. These parameters describe the EM properties of a dielectric box placed on the C-section*

6.4.2.1. *Issue*

Very high stresses weigh on the manufacture of the large antenna reflectors used on satellites to limit deformations. This results in extremely high costs. One way to reduce these is to precisely measure the deformation (with an error of less than 1 mm) of the reflectors in flight in order to compensate for it during the formation of the beams emitted. In such cases, antenna dishes can be produced using less specific materials (that is to say, less expensive).

6.4.2.2. *Latest developments and solution implemented*

In the literature, and more particularly with regard to RFID, few approaches have been introduced for measuring the deformation of objects. The approach described in [CAI 12] uses a network of passive UHF RFID tags arranged on the object we are seeking to monitor. Information on deformation is obtained by measuring the minimum activation power of each tag. Precision on the order of a dozen centimeters has been obtained. In order to increase the precision, the use of UWB pulses coupled with a chipless approach seems to be a well-adapted solution. For example, the UWB RFID localization approach introduced in [CHO 09] offers good precision when point-to-point distances are measured. However, very few localization projects of this type have addressed the issue of measuring variations of several distances at the same time, variations of less than 1 mm, in order to monitor in real time the deformation of an object located several meters away from the reader. Chipless RFID technology as presented in this book has several advantages as a solution to this issue. By doing away with the chip, and thus with the imprecisions and measurement errors associated with it, as well as the delay inherent in the protocols it must manage, precision higher than that obtained with classic RFID can be attained. The approach introduced consists of a system composed of a chipless RFID reader carefully positioned in relation to the reflector, and of chipless tags placed on the reflector at the places where measurement of deformation is desired. Information on deformation will be obtained from the measurement of time-in-flight. An anticollision system based on a distinction in frequency will make it possible to link the variation detected to the shifting of one of the tags present on the reflector

[PER 14]. It is a matter, then, of measuring distances and extracting a deformation from them. Unlike [CAI 12], it is possible here to obtain the desired information by sending a single pulse.

6.4.2.3. *Principle of methodology*

The basic idea of the solution proposed relies on the use of a chipless tags grid (where the tags are positioned at different places on the reflector), and on a reader that generates a questioning signal and measures the signals reflected by the tags, as shown in Figure 6.9. The tag's response is modified by any deformation/shift in relation to an initial position. In order to make a link between the distance information obtained and the corresponding reflector area, several approaches – that is, a temporal approach or a frequential one – can be used. In the case of the temporal approach, as with SAW tags, each tag will be associated with a specific temporal range. The distance variation will be obtained by recovering the information on the temporal gap in the signal coming from the tag, compared to the temporal range allocated to it.

The same thing can be done with a temporal approach; this time, a frequential window will be associated with each tag. In this case, the tag-sensor can simply consist of an element resonating at a well-defined frequency [VEN 13]. A wideband excitation (UWB pulse) sent by the reader illuminates all the tags positioned on the reflector. The signal backscattered by the tag-and-reflector set contains information on all the tag positions. It is then up to the reader to make a distinction between all this information; the procedure used is described in Figure 6.10. It consists of applying a Fourier transform to the temporal signal reflected by the group of tags. Next, a set of frequential windows is created in order to isolate each response from the tags (tag number i, for example). The opposite Fourier transform is then applied on this filtered signal to return to the temporal domain. The information in which we are interested is now contained in the peak of this signal. Based on this, the antenna-tag number i distance (and thus the deformation) is accessible both in the amplitude and in the temporal position t_i of this maximum. Note that it is also possible to extract the distance by using the phase of the backscattered signal expressed in frequency.

Let us take an example based on the frequential approach. We are interested in evaluating the performances of this approach using chipless tags in the form of inverted double Ls operating in cross-polarization (see Figure 6.11) [VEN 13].

Figure 6.9. *Operating schema of the approach used to measure the deformation of a dish antenna*

6.4.2.4. *Experimental results*

To validate the approach in a favorable case, a network analyzer (HP 8720D) has been used to act as a "reader". Figure 6.12 shows the measurement bench used. The gain of the antennas used is on the order of 12 dBi between 1.5 and 6 GHz. The VNA output power is 0 dBm. The two inverted double-L tags with a resonant frequency of 4.72 GHz are placed on a support attached to a micrometric positioning sliding plate, at a distance d_0 equal to 35.5 cm of the horn antenna. The smallest movements allowed are 20 µm, and the total translation range is 15 mm. The procedure previously described is then implemented; Figure 6.13 shows the measurement signals once processed for 11 tag movements in relation to the antenna (with a constant step of 1 mm). We can see a temporal shift and a variation in the amplitude of the signals. When the tag moves farther away from the antenna, the amplitude decreases, and the time corresponding to the signal maximum is delayed.

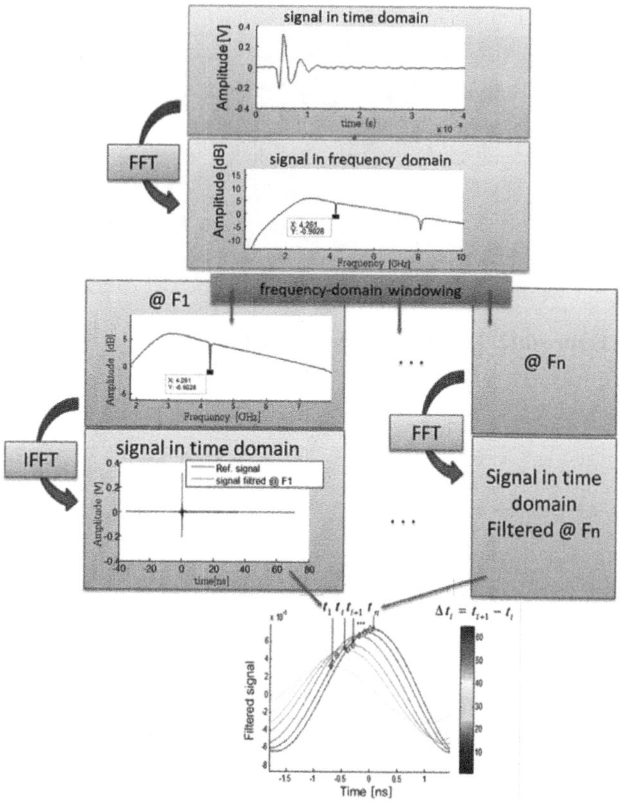

Figure 6.10. *Descriptive schema of the frequential approach applied to measure a deformation with chipless tags. For a color version of this figure, see www.iste.co.uk/perret/radio.zip*

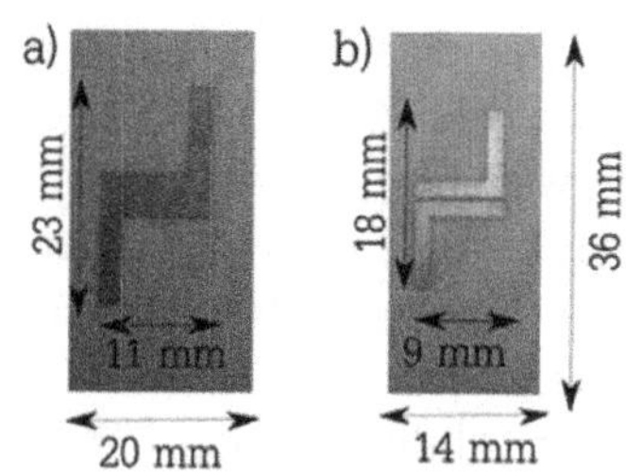

Figure 6.11. *Photos of the two inverted double-L tags used for the experiment. a) The first one (tag 1) resonates at 4.72 GHz and b) the second (tag 2) resonates at 5.26 GHz*

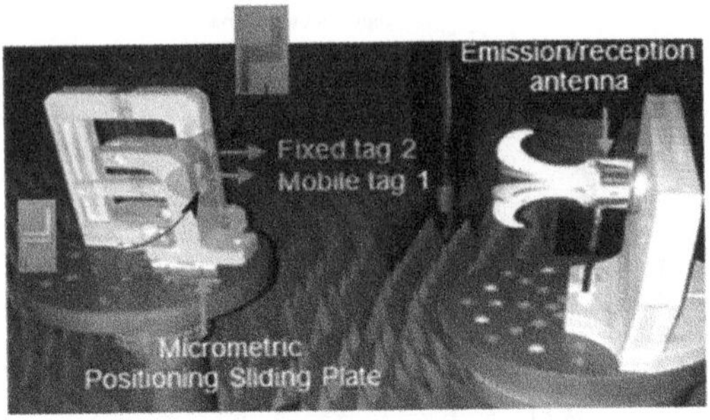

Figure 6.12. *Measurement bench used in an anechoic chamber for deformation measurement*

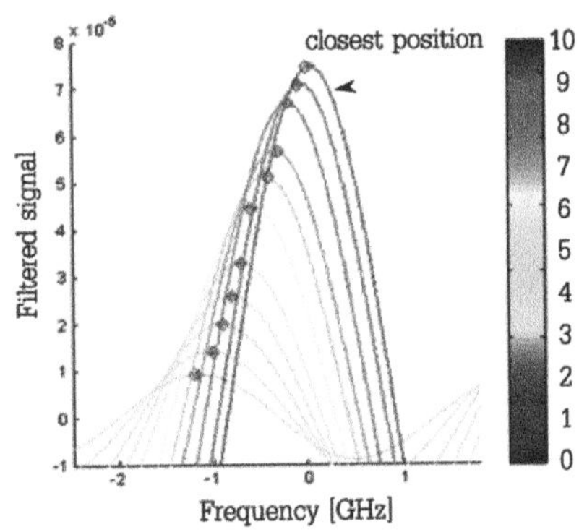

Figure 6.13. *Signals reflected and then processed (see Figure 6.10) from the two inverted double-L tag at 4.72 GHz for 11 movements of 1 mm each. For a color version of this figure, see www.iste.co.uk/perret/radio.zip*

The movement measurements were taken in the presence of several tags located in the reader field. As Figure 6.12 shows, two tags facing the reader and 30 cm away from it have been positioned. Next, only one tag was moved at a time, with the other remaining still. The movement distance is 0.5 mm per step. For configuration 1, the tag with a frequency of resonance of about 4.72 GHz is placed on the micrometric positioning sliding plate. The tag functioning at 5.26 GHz is static. Configuration 2 consists of inverting the two tags, that is, placing the tag at 5.26 GHz on the mobile support. With each movement, the RCS of the tags is measured.

After filtering around the resonance frequency of the tag studied, and using an inverse Fourier transform (as previously explained), we obtain Figure 6.14, the postprocessing temporal response, where information on distance is present. We can see that with a movement step of 0.5 mm, the curves are clearly separated both in amplitude and time. It is thus possible to link these variations and movements of 0.5 mm. For example, detection of the time corresponding to the maxima of each curve gives the temporal shift between the different positions of the tags. From this, we can deduce a value that is a direct function of the movement. A calibration technique is used to determine the deformation undergone by the reflector. Figure 6.14(c) shows the filtered signals around the frequency corresponding to the stationary tag. We can see that the maxima are obtained at almost the same time. However, there is a variation in amplitude linked to the movement of the tag at 5.26 GHz. It appears preferable, then, to use the temporal variation to determine movement. Note that, despite the presence of several tags in the reader field, the method of estimation introduced can still be used to detect variations of at least 0.5 mm in one of the tags.

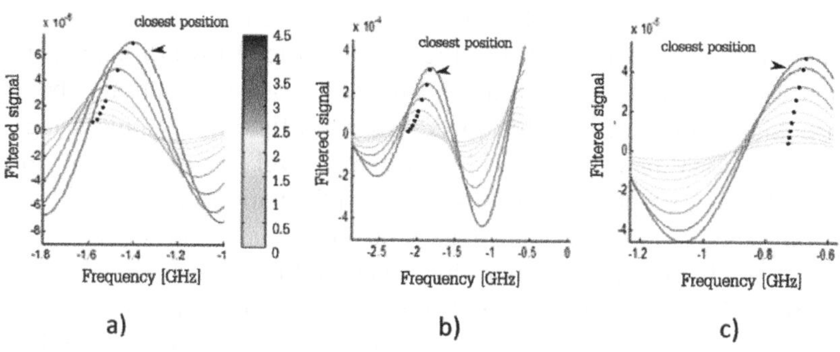

Figure 6.14. *Results of measurements of postprocessed reflected signals: a) tag 1 (signal filtered at 4.72 GHz), configuration 1; b) tag 2 (signal filtered at 5.26 GHz), configuration 2; and c) tag 1 (signal filtered at 4.72 GHz), configuration 2. The color bar indicates, for the three figures, the value in mm of the 10 movements made (steps of 0.5 mm). The tags are moving toward the reader. For a color version of this figure, see www.iste.co.uk/perret/radio.zip*

In conclusion, the measurement of the distance variation of a set of chipless RFID tags using time-of-flight gives satisfactory results and shows the reliability of the approach introduced. The use of several tags in the same reading zone, each resonating at different frequencies, allows us to determine their variations in position. It is, therefore, possible to distribute a certain number of tags on a surface so as to monitor its deformation.

A final eagerly-awaited functionality in chipless RFID is the possibility of being able to reconfigure the information present on a tag once this tag is produced. The main difficulty here is that the approach must remain compliant with the idea of chipless, that is, potentially printable and low-cost. We will now present both the current results and the research perspectives related to this theme, which, as we will see, moves the chipless problem.

6.5. *Reconfigurable chipless*

The objective has been to validate the concept of a new family of chipless tags in which the identifier can be reconfigured while the tag is in use. In terms of strategic positioning, in order to be competitive with the barcode, chipless RFID must offer more functionalities to the user. Compared to UHF RFID, the idea is to tend toward almost the same functionalities but at a much lower tag cost. It would also be of particular interest to be able to design reconfigurable chipless tags. This functionality would provide a decisive level of flexibility of use; it would then be possible to modify the information contained in the tag at will. A tag could have its data erased and subsequently be reused. It should be noted that this functionality is incompatible with the use of barcodes (single-use, disposable tags); thus, it is particularly relevant to supply proof-of-concept for a new generation of chipless tags whose information would be reconfigurable. The tags in question would have a write and rewrite function, which until now have been reserved for chipped RFID. This is an original concept, and is made possible by the development of new microelectronic technologies. The challenge here is to be able eventually to produce these tags with dielectric materials at low cost, and even with current printing techniques. To do this, research works have begun with rudimentary chipless tags, to which we are attempting to add a "rewrite" function using controllable RF switches.

6.5.1. *Operating principle of CBRAM*

The reconfigurable part is created using "Conductive-Bridging RAM" (CBRAM) technology as a basis [AKI 10, DER 10, KUN 05, VAL 11]. This technique is used to produce programmable resistive elements that preserve their state even in the absence of a power supply (non-volatile behavior). They are based on a metal-insulator-metal (MIM) structure. Specific pairs of metals (e.g. copper and aluminum) are used, as well as a large number of dielectrics (oxides [BER 11], polymers [POT 79] and chalcogenides) used to obtain the function desired. The application of voltage between the two metals results in the creation of conductive filaments that end by connecting the two electrodes and thus causing the structure to switch; this puts the device in "ON" mode, a state that is preserved even when polarization of the structure is stopped. To change it, all that is required is the application of a reverse voltage polarization, which causes the filaments to break. This "OFF" mode is then preserved until the direction of polarization is reversed, reforming the filaments and shifting back to "ON" mode. Note that only the very first switch into "ON" mode differs fairly significantly from the other mode changes; it is effectively at that moment that Cu + ions, for example, migrate into the dielectric in order to create the filament to be reduced with electrons derived from the inert metal. This stage generally requires voltage higher than the one applied for subsequent switching, in which only a part of the filaments needs to be reformed in order to create the conductive bond. For applications related to microelectronics, it has been demonstrated that it is possible to cause the equivalent resistance value of the structure to vary by playing on the time or voltage applied. In this case, we are in a position to increase the diameter of the filaments, as well as their number, in order to obtain resistance values on the order of a dozen ohms [LIU 09, RUS 09]. Impedance values of several mega-ohms are obtained in the "OFF" mode. These structures are currently being studied for the production of non-volatile CBRAM rapid access memory [KUN 05]. In comparison to these projects, the objective here is to use the CBRAM principle to produce RF switches. Another variant, called a programmable metallization cell (PMC) and having two different states of conductivity (based on phase-change material), is also being investigated for the production of memories and RF switches [CRU 14, YON 13].

We will note here the significant divergences that exist in terms of specifications and desired performance between memories and RF switches which are very different applications from an applicative point of view. In the case of memories, consumption and time of switching/memory access are determining factors, while in RF it is above all frequency behavior and particularly the transmission ratio between ON and OFF modes that is decisive. More precisely, the merit factor of a switcher is related to the product of the equivalent resistance in ON mode (R_{On}) multiplied by the equivalent capacity in OFF mode (C_{Off}). In this case, we are seeking to minimize this product as far as possible. Given the specific characteristics proper to each of these applications, the principal differences between memories and the RF switch version can be found in the dimensions of structures. In RF, we must ensure the lowest C_{Off} capacity possible. We must also seek configurations in which the dielectric is as thick as possible, that is, it must absolutely be significantly thicker than 50 nm in general for memories. Studies have shown that it is possible to execute switching with thicknesses of several hundred nanometers [VEN 12d].

Note, however, that the earliest projects began at low frequency on CBRAM technology for the production of reconfigurable analog circuits [PER 10a], such as tunable filters [DRI 10]. A project closer in nature to the function desired for chipless tags has shown the possibility of producing an RF switch operating between 1 and 6 GHz [NES 10, NES 08]. A comparison with switches based on classic approaches is presented in [NES 10], and the results are spectacular. However, unlike the work carried out by Arnaud Vena [VEN 12d], this research focused on the use of a fairly specific dielectric ($GeSe_2$ glass doped with silver atoms). To provide proof-of-concept for the integration of RF switches using CBRAM technology in a chipless RFID tag, the use of common, potentially printable materials is expected.

If we look very briefly at the latest developments in integrated RF switches, RF MEMS offer very good performances compared to other technologies based on FET transistors or PIN diodes. For classic applications (not including spatial specifications), with the exception of switching time and power level, MEMS yield better performances in terms of insertion losses (0.05–0.2 dB), isolation (30 dB), cutoff frequency (20–80 THz) and power consumption (0.05–0.1 mW). This results in series capacitances in OFF mode (C_{Off}) of between 1 and 10 fF, as well as contact resistances in ON mode (R_{On}) of between 0.1 and 2 ohms. However, this technology requires complex production techniques using expensive components, and control voltages are significant, ranging from 5 to 80 V, which is higher than in the FET and PIN approaches (3–5 V). In the course of the development of electronics that are ever more integrated, low-loss, linear and rapid, it is therefore legitimate to try to evaluate the potential of the CBRAM approach (as well as PMC approach) applied to hyperfrequencies.

6.5.2. *Example of a reconfigurable chipless tag*

The CBRAM structure is extremely simple. Figure 6.15 shows an example of the implantation of this technology in a chipless tag in order to make it reconfigurable. The RF switch is used to modify the geometric length of the tag and thus its resonance frequency. The principal coding approaches previously introduced, as well as the topologies of chipless tags, can be reused. All that is required is to carefully position the switches to obtain the desired RF signatures and thus the different configurations. As an example, in Figure 6.15(a), the arrangement of switches at the center of the dipoles makes it possible to obtain basic OOK coding in frequency. We can see in Figure 6.15(b) how the modifiable element, the CBRAM cell, could be integrated. Thanks to the $Cu/SiO_2/Al$ layers, filaments inside the SiO_2 dielectric can be created to produce a short circuit. In terms of realization, if we start from a classical chipless tag (that is, a substrate with a conductive pattern), we must begin by depositing the dielectric layer, being careful to leave a copper access area on the arm that does not contain the MIM structure. Then, we must deposit aluminum in such a way that we can cover part of the arm of the antenna, thus creating the MIM stacking and ensuring electrical contact with the other arm. This results in a horizontal stacking of layers (out-of-plane RF switch). It is also possible to create a vertical MIM structure (in-line RF switch), but the thinness of the dielectric (typically on the order of a few hundred nanometers) makes this second configuration more difficult to realize in practice. The objective has been to demonstrate the feasibility of this type of approach applied to RF, that is, with different materials and larger dimensions than the ones classically used for memories. In addition, classic RF substrates (RO4003C and FR4) have been used as supports while also contributing to the first metallization of the copper (at a thickness of 17 μm).

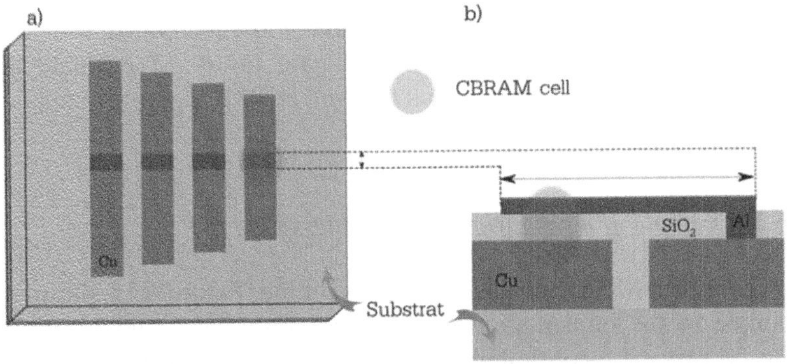

Figure 6.15. *Reconfigurable chipless tag with three dipole resonators. a) View from above and b) cross-section of the CBRAM cell. For a color version of this figure, see www.iste.co.uk/perret/radio.zip*

Next, to create the dielectric layer, various solutions have been tested, such as the deposit of SiO_2 by cathodic pulverization, or the deposit of resins (PMMA) via spin coating to preserve simple production techniques. The thicknesses of the layers tested measured between 160 nm and 1 μm, depending on the dielectrics. Then, the aluminum layer was deposited via evaporation (at a thickness of 500 nm per deposit) using a metallic mask with a large number of circular holes. In this way, structures similar to those shown in Figure 6.16(b) have been obtained. The method of RF characterization is shown in Figures 6.16(a) and (c). The DC properties of the switch produced have been determined using a DC generator, a voltmeter and an ammeter connected to a personal computer (PC) to monitor the programming cycle.

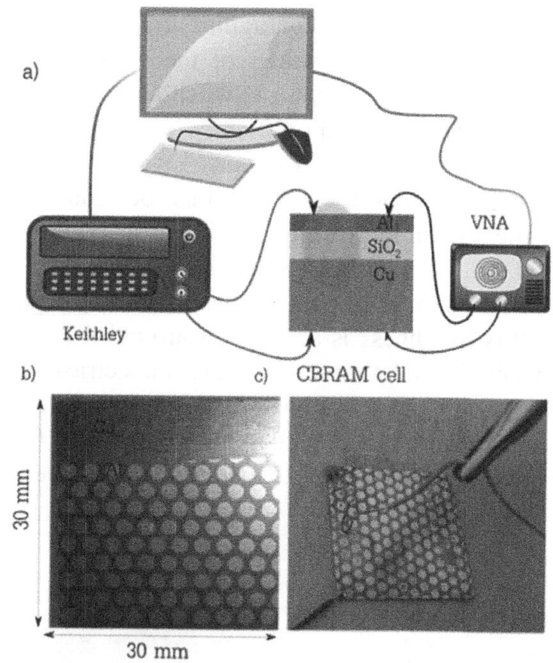

Figure 6.16. *a) Operating schema of DC and RF measurement. b) Photograph of samples produced, with each aluminum disk corresponding to an RF switch, that is, to an MIM stack. c) View of the measurement bench used to measure the transmission coefficients of the switch using a VNA*

The principal result of this preliminary study is that it is quite possible to induce switching in this type of structure. With several volts (from 1 to 30 V, depending on the dielectrics and thicknesses used), an ON mode can be obtained, and a blocked mode can be achieved by reversing the voltage. Current-voltage characterizations (I-V curves) have shown that this schema can be reproduced in a large number of cycles. Although they were not designed for it, these structures have also been characterized in RF (see Figure 6.16(c)), and the S parameters are measured for the two modes of the switch. The results obtained for the structure with SiO_2 are presented in Figures 6.17(a) and (b); in this case, no de-embedding/calibration was used. We can, however, see good isolation levels (–20 dB for OFF mode) up to 400 MHz and insertion losses of 1.66 dB for ON mode. Resistances of a few ohms have been observed in certain structures in

ON mode. These results are encouraging, and these structures have significant potential. Once the values of the equivalent components were determined via measurements, a switch able to function at up to 10 GHz was simulated. The main difficulty lies in realization, since in order to reduce the inherent capacitance of the MIM device, the surface area of the aluminum electrode must be reduced. The device obtained and the simulation results are shown in Figures 6.17(c) and (d). Compared to the samples tested, which were as common as possible, the main limitation lies in the extreme roughness of the copper layer; this roughness is actually of around the same thickness (or even thicker in most cases) as the dielectrics deposited. In addition, few of the switches are functional on the same plate. However, the results obtained suggest that the use of CBRAM cells may enable chipless technology to undergo a new period of rapid expansion.

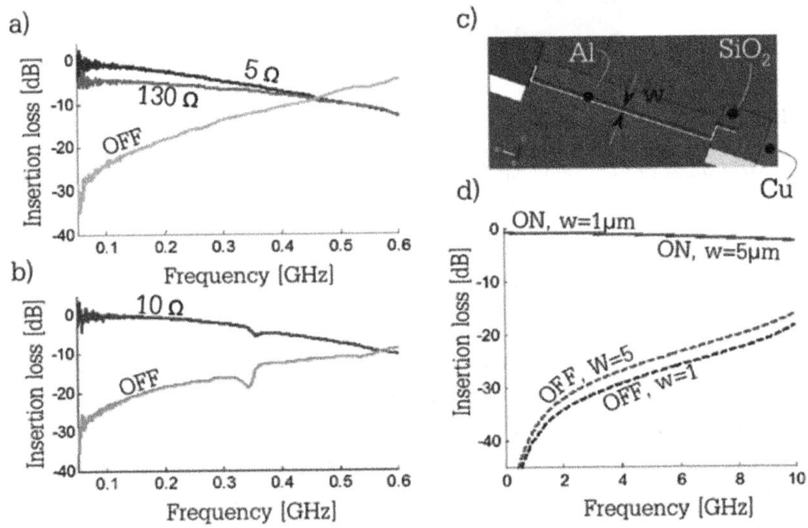

Figure 6.17. *Insertion losses measured for the sample shown in Figure 6.16(b), with a SiO_2 thickness of a) 160 nm and b) 320 nm. c) Geometry of switch designed to function at up to 10 GHz. d) Simulated insertion losses for the switch presented for several slope lengths w corresponding to the aluminum electrode. The SiO_2 thickness is 500 nm*

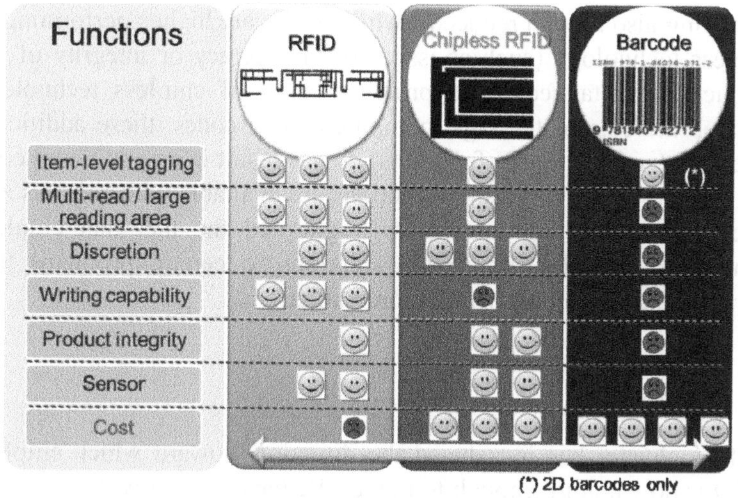

Figure 6.18. *Positioning of chipless RFID in relation to barcodes and classic RFID. Comparison of principal functionalities available, notably the sensor function, for each technology*

Before ending this chapter, it is of importance in light of these results to compare this technology to the RFID and barcode identification approaches. A summary of this comparison is shown in Figure 6.18. It is notable that in terms of functionalities, there is a heavy argument to be made in favor of RFID; the only black mark on the table has to do with price. Barcodes offer very few functions, but the technology is proved, highly widespread, referential and extremely low-cost. If we consider new uses related to the proliferation of smartphones and thus the cameras they contain, barcodes – which rely on imagery – will pass the test without difficulty, and even with flying colors. Today, every smartphone user can read a barcode at will. RFID cannot easily match this, despite the presence of near-field communication (NFC). Without exception, NFC functioning at 13.56 MHz fails to solve the problem, since for the moment this solution is not compatible with UHF tags, and is thus incompatible with a large part of the traceability market. Chipless technology also offers a good argument in terms of functionalities; some of these tags are downgraded versions of what RFID can do (in terms of the possibility of carrying out multiple readings; distance and flexibility of reading also remain reduced), while others are higher-performing in chipless technology (such as issues of tag secrecy or integrity of the product being tagged). The principal asset of chipless technology remains the cost of the tags: compared to barcodes, these additional functions (such as sensor functions) do not result in an additional cost, and some of RFID's flagship functions are available with chipless at a much lower cost. Moreover, we might add the fact that chipless technology should also provide solutions to certain problems that currently exist in the authentication market.

6.6. Conclusion

This chapter has introduced the directions toward which chipless RFID is now mature enough to move. We have seen how it is possible to imagine the design of a new family of chipless tag-sensors with an identifier, communicating via radiowaves and able to be produced at low cost. All of these developments are based on the principle and use of chipless RFID technology, with the desire to continue using simple devices that are potentially printable and low cost. The example of the tag-sensor has shown that, via the addition of materials such as silicon nanowires that are sensitive to certain physical values, it is possible to move toward a sensor function.

With regard to switches, the study presented involves the transposition into the domain of RF of microelectronic technology used in creating the memories of tomorrow. From there, we can imagine the perfection of RF switches based on this recent approach that are potentially compatible with printing technologies using conductive ink, which should result in a low unit cost. Switches are basic RF components used to reconfigure devices during use to make them more agile. It has been shown how to use these switches to make printed chipless tags reconfigurable. These devices do not exist at present, and would in fact represent a significant technological break, making it possible to produce blank tags for subsequent encoding, or simply to erase tags in order to reuse them. This is an innovative concept, made possible by a combination of chipless technology and the development of new microelectronic technologies.

We have shown in this chapter that chipless RFID has many assets, making it possible to offer many highly sought-after functionalities besides identification. The fields of application are immense. The perfection of extremely low-cost tag-sensors is currently very desirable; as an example, there is a demand to be able to create processes to track materials throughout stages of manufacturing. Low-cost tag-sensors are necessary to do this, but they must often also be highly resistant to the various production procedures implemented. In these examples functioning in closed loops, chipless should provide a highly effective customized solution. It is true that the results presented in this chapter remain highly prospective and need to be completed by more advanced studies. We have considered perspectives on the chipless approach here, rather than available key-in-hand solutions. However, even the perspectives are significant.

6.7. Bibliography

[AKI 10] AKINAGA H., SHIMA H., "Resistive random access memory (ReRAM) based on metal oxides", *Proceedings of the IEEE*, vol. 98, no. 12, pp. 2237–2251, 2010.

[AUB 09] AUBERT H., CHEBILA F., JATLAOUI M.M., *et al.*, Dispositif de mesure comprenant un diffuseur électromagnétique, Patent no. WO 2010136388 (A1), 2009.

[BAL 08] BALACHANDRAN M., SHRESTHA S., AGARWAL M., *et al.*, "SnO_2 capacitive sensor integrated with microstrip patch antenna for passive wireless detection of ethylene gas", *Electronics Letters*, vol. 44, no. 7, pp. 464–466, 2008.

[BEL 13] BEL-KAMEL E., PERRET E., VENA A., *et al.*, "Utilisation de la technologie RFID chipless pour la mesure de déformation de réflecteurs", *18èmes Journées Nationales Microondes*, Paris, France, 2013.

[BER 11] BERNARD Y., RENARD V., GONON P., *et al.*, "Back-end-of-line compatible conductive bridging RAM based on Cu and SiO_2", *Microelectronic Engineering*, vol. 88, no. 5, pp. 814–816, May 2011.

[CAI 11a] CAIZZONE S., MARROCCO G., "RFID grids: part II – experimentations", *IEEE Transactions on Antennas*, vol. 59, no. 8, pp. 2896–2904, 2011.

[CAI 11b] CAIZZONE S., MARROCCO G., "RFID Grids: Part II – Experimentations", *IEEE Transactions on Antennas and Propagation*, vol. 59, pp. 2896–2904, 2011.

[CAI 12] CAIZZONE S., MARROCCO G., "RFID-grids for deformation sensing", *IEEE International Conference on RFID*, Orlando, Florida, 2012.

[CHE 11] CHEN X., ZHANG J., WANG Z., *et al.*, "Humidity sensing behavior of silicon nanowires with hexamethyldisilazane modification", *Sensors and Actuators B: Chemical*, vol. 156, no. 2, pp. 631–636, 2011.

[CHO 09] CHOI J.S., LEE H., ELMASRI R., *et al.*, "Localization systems using passive UHF RFID", *5th International Joint Conference on INC*, IMS and IDC, 2009.

[CHU 11] CHUAN-PO W., CHIEN-WEI L., CHIE G., "Silicon nanowire temperature sensor and its characteristic", *IEEE International Conference on Nano/Micro Engineered and Molecular Systems (NEMS)*, 2011.

[CRU 14] CRUNTEANU A., MENNAI A., GUINES C., *et al.*, "Out-of-plane and inline RF switches based on Ge2Sb2Te5 phase-change material", *IEEE MTT-S International Microwave Symposium (IMS)*, Tampa Bay, Florida, 2014.

[CUI 01] CUI Y., WEI Q., PARK H., *et al.*, "Nanowire nanosensors for highly sensitive and selective detection of biological and chemical species", *Science,* vol. 293, pp. 1289–1292, 2001.

[DAI 14] DAIKI M., PERRET E., TEDJINI S., "Design of near field UHF RFID reader antenna integrated into clothing", *IEEE International Conference on RFID-Technology and Applications*, Tampere, Finland, pp. 261–265, 2014.

[DER 10] DERHACOBIAN N., HOLLMER S.C., GILBERT N., *et al.*, "Power and energy perspectives of nonvolatile memory technologies", *Proceedings of the IEEE*, vol. 98, no. 2, pp. 283–298, 2010.

[DRI 10] DRISCOLL T., QUINN J., KLEIN S., *et al.*, "Memristive adaptive filters", *Applied Physics Letters*, vol. 97, no. 9, pp. 093502–093502-3, 2010.

[GUI 12] GUILLET A., VENA A., PERRET E., *et al.*, "Design of a chipless RFID sensor for water level detection", *15th International Symposium on Antenna Technology and Applied Electromagnetics (ANTEM)*, Toulouse, France, 2012.

[JAT 07] JATLAOUI M.M., PONS P., AUBERT H., "Radio frequency pressure transducer", *2007 European Microwave Conference*, pp. 736–739, 9–12 October 2007.

[JAT 10] JATLAOUI M.M., CHEBILA F., BOUAZIZ S., *et al.*, "Original identification technique of passive EM sensors using loaded transmission delay lines", *2010 European Microwave Conference (EuMC)*, pp. 1106–1109, 28–30 September 2010.

[KUN 05] KUND M., BEITEL G., PINNOW C.U., *et al.*, "Conductive bridging RAM (CBRAM): an emerging non-volatile memory technology scalable to sub 20nm", *IEDM Technical Digest, IEEE International Electron Devices Meeting*, pp. 754–757, 5 December 2005.

[LI 09a] LI H., ZHANG J., TAO B., *et al.*, "Investigation of capacitive humidity sensing behavior of silicon nanowires", *Elsevier*, vol. 41, no. 4, pp. 600–604, February 2009.

[LI 09b] LI Y., RONGWEI Z., STAICULESCU D., *et al.*, "A novel conformal RFID-enabled module utilizing inkjet-printed antennas and carbon nanotubes for gas-detection applications", *IEEE Antennas and Wireless Propagation Letters*, vol. 8, pp. 653–656, 2009.

[LIE 09] LIEBERZEIT P., PALFINGER C., DICKERT F., *et al.*, "SAW RFID-tags for mass-sensitive detection of humidity and vapors", *Sensors*, vol. 9, pp. 9805–9815, 2009.

[LIU 09] LIU Q., DOU C., WANG Y., *et al.*, "Formation of multiple conductive filaments in the Cu/ZrO_2: Cu/Pt device", *Applied Physics Letters*, vol. 95, no. 2, pp. 023501–023503, 2009.

[MAR 11] MARROCCO G., "RFID grids: part I – electromagnetic theory", *IEEE Transactions on Antennas and Propagation*, vol. 59, no. 3, pp. 1019–1026, 2011.

[MCG 06] MCGRATH M.P., PHAM A., "Carbon nanotube based microwave resonator gas sensors", *International Journal of High Speed Electronics and Systems*, vol. 16, no. 4, pp. 913–935, 2006.

[MET 14] METEO FRANCE, "La mesure de l'humidité", Available at http://educationmeteofrance.com/education/mesures/humidite?document_id=21875&portlet_id=58718.

[NAI 11] NAIR R., PERRET E., TEDJINI S., "Chipless RFID based on group delay encoding", *IEEE International Conference on RFID Technologies and Applications (RFID-TA)*, Barcelona, Spain, 2011.

[NAI 13a] NAIR R.S., PERRET E., TEDJINI S., "Group delay modulation for pulse position coding based on periodically coupled C-sections", *Annals of Telecommunications*, Special Issue on Chipless RFID, vol. 68, no. 7–8, pp. 447–457, 2013.

[NAI 13b] NAIR R.S., PERRET E., TEDJINI S., "A temporal multi-frequency encoding technique for chipless RFID based on C-sections", *Progress in Electromagnetics Research B*, vol. 49, pp. 107–127, 2013.

[NES 08] NESSEL J.A., LEE R.Q., MUELLER C.H., *et al.*, "A novel nanoionics-based switch for microwave applications", *IEEE MTT-S International Microwave Symposium Digest*, pp. 1051–1054, 15–20 June 2008.

[NES 10] NESSEL J., LEE R., Chalcogenide nanoionic-based radio frequency switch, US Patent No. 7.923.715 B2, 2010.

[NIK 07] NIKITIN P.V., RAO K.V.S., LAZAR S., "An overview of near field UHF RFID", *IEEE International Conference on RFID*, pp. 167–174, 26–28 March 2007.

[NOM 94] NOMURA T., OOFUCHI K., YASUDA T., *et al.*, "SAW humidity sensor using dielectric hygroscopic polymer film", *Proceedings of the IEEE Ultrasonics Symposium*, vol. 1, pp. 503–506, 31 October–3 November 1994.

[PEN 09] PENG K.-Q., WANG X., LEE S.-T., "Gas sensing properties of single crystalline porous silicon nanowires", *Applied Physics Letters*, vol. 95, no. 24, Article id. 243112, 2009.

[PER 10a] PERSHIN Y.V., DI VENTRA M., "Practical approach to programmable analog circuits with memristors", *IEEE Transactions on Circuits and Systems I: Regular Papers*, vol. 57, no. 8, pp. 1857–1864, 2010.

[PER 10b] PERADOVIC S., KARMAKAR N.C., "Chipless RFID tag with integrated sensor", *Proceedings of the IEEE International Conference in Sensors*, pp. 1277–1281, 1–4 November 2010.

[PER 14] PERRET E., NAIR R.S., KAMEL E.B., *et al.*, "Chipless RFID tags for passive wireless sensor grids", *2014 XXXIth URSI General Assembly and Scientific Symposium (URSI GASS)*, Beijing, China, 2014.

[POT 79] POTEMBER R., POEHLER T., COWAN D., "Electrical switching and memory phenomena in CuTCNQ thin films", *Applied Physics Letters*, vol. 34, no. 6, pp. 405–407, 1979.

[RUS 09] RUSSO U., KAMALANATHAN D., IELMINI D., *et al.*, "Study of multilevel programming in programmable metallization cell (PMC) memory", *IEEE Transactions on Electron Devices*, vol. 56, no. 5, pp. 1040–1047, 2009.

[SHE 11] SHENG L., DAJING C., YUQUAN C., "A surface acoustic wave humidity sensor with high sensitivity based on electrospun MWCNT/Nafion nanofiber films", *Nanotechnology*, vol. 22, no. 26, 2011.

[SHR 09] SHRESTHA S., BALACHANDRAN M., AGARWAL M., *et al.*, "A chipless RFID sensor system for cyber centric monitoring applications", *IEEE Transactions on Microwave Theory and Techniques*, vol. 57, no. 5, pp. 1303–1309, 2009.

[SKU 10] SKUCHA K., FAN Z., JEON K., *et al.*, "Palladium/silicon nanowire Schottky barrier-based hydrogen sensors", *Sensors and Actuators B: Chemical*, vol. 145, pp. 232–238, 2010.

[STA 06] STAR A., TU E., NIEMANN J., *et al.*, "Label-free detection of DNA hybridization using carbon nanotube network field-effect transistors", *National Academy of Sciences*, vol. 103, no. 4, pp. 921–926, 2006.

[TED 11] TEDJINI S., PERRET E., VENA A., *et al.*, "Sensing properties of chipless RFID tags based on nanomaterials", *Workshop on Nanotechnology-Enabled RF and Cognitive Devices, Components and Systems (IEEE-IMS '11)*, Baltimore, June 2011.

[THA 10] THAI T.T., CHEBILA F., MEHDI J.M., *et al.*, "Design and development of a millimetre-wave novel passive ultrasensitive temperature transducer for remote sensing and identification", *2010 European Microwave Conference (EuMC)*, pp. 45–48, 28–30 September 2010.

[VAL 11] VALOV I., WASER R., JAMESON J.R., *et al.*, "Electrochemical metallization memories – fundamentals, applications, prospects", *Nanotechnology*, vol. 22, no. 25, 2011.

[VEN 12a] VENA A., Contribution au développement de la technologie RFID sans puce à haute capacité de codagel, PhD Thesis, University of Grenoble, Valence, France, 2012.

[VEN 12b] VENA A., PERRET E., MANNEQUIN C., *et al.*, "Conception d'un switch RF à base de filaments conducteurs nanométriques", *12ème édition des Journées de Caractérisation Microondes et Matériaux (JCMM '12)*, Chambéry, France, 2012.

[VEN 12c] VENA A., PERRET E., TEDJINI S., *et al.*, "A compact chipless RFID tag with environment sensing capability", *IEEE MTT-S International Microwave Symposium Digest*, Montréal, Canada, 2012.

[VEN 12d] VENA A., PERRET E., TEDJINI S., *et al.*, "A fully passive RF switch based on nanometric conductive bridge", *IEEE MTT-S International Microwave Symposium (IMS)*, Montréal, Canada, 2012.

[VEN 13] VENA A., PERRET E., TEDJINI S., "A depolarizing chipless RFID tag for robust detection and its FCC compliant UWB reading system", *IEEE Transactions on Microwave Theory and Techniques,* vol. 61, no. 8, pp. 2982–2994, August 2013.

[YON 13] YONGHYUN S., HUMMEL G., RAIS-ZADEH M., "RF switches using phase change materials", *IEEE 26th International Conference on Micro Electro Mechanical Systems (MEMS)*, 2013.

[YOO 05] YOON H., XIE J., ABRAHAM J.K., *et al.*, "Passive wireless sensors using electrical transition of carbon nanotube junctions in polymer matrix", *Smart Materials and Structures*, vol. 15, no. 1, pp. S14–S20, 2005.

Conclusion

With regard to passive ultra high frequency (UHF) radio frequency identification (RFID), we have seen that the issue of the antenna is truly at the heart of many of this technology's limitations. A design approach based on the combined use of different softwares has shown that it is possible to gain in reading flexibility simply by working with the antenna's topology; that is, by moving toward new forms of antennas. To do this, electromagnetic (EM) simulation software has been coupled with calculation software. The main motivation behind the integration of these two types of softwares is to take advantage of data manipulation, the very wide range of functions (optimization functions, for example), and the extended graphic possibilities of Matlab, while benefiting from a high-performance, multifunction EM simulator. The design procedure for RFID antennas is original and makes it possible to break with the empirical approach classically employed. Moreover, the integration of these different simulators is a general approach. The resulting application is fully adjustable and can be used in any area of EM circuit design.

Given that RFID tags are by nature very sensitive to the environment (especially if nothing particular is done in terms of antenna design), it seems logical to use this behavior to obtain information about the tag's environment. This entails the addition of a sensor function to the unique identifier (ID) of the tag. All of this relies on the impedance mismatch of the antenna; this variation will occur directly on the signal backscattered by the tag. We have also seen that the RFID approach can be adapted to the characterization of antennas with small dimensions. In fact, in this case radar cross-section (RCS) measurements can replace classic measurements using cables. Two of the three impedance states necessary to extract the information can be obtained directly by connecting an RFID chip to the antenna. If necessary, the final measurement can be taken by removing the chip, leaving the antenna in open circuit. Even if the performances remain to be validated in practice, this approach is still promising, all the more so, because some chips can function on very wide frequency bands of between 13.56 MHz and 2.45 GHz.

We saw in Part 1 of this book that RFID, based on the transmission of radio frequency (RF) waves, yields a very rich functional range; the acquisition of the ID is greatly facilitated; we can achieve volumetric reading and tags containing variable information, and all of this in a highly discreet manner. These functions are impossible to implement with barcodes, and we might say that the latter have been irretrievably outstripped; yet, if we look at the figures, this is absolutely not the case. In fact, 70% of the objects manufactured in the world are equipped with barcodes; that is, around 15,000 billion units per year. The success of this technology is easy to explain; it works very well and is extremely low-cost. Based on these observations, we can see that a technology able to combine the flexibility of use of RF waves and the extremely low cost of barcodes would have a bright future. We believe that the solution is none other than chipless RFID; that is, a paper tag, for example, with an appearance very similar to a barcode, with the difference being that it is printed with conductive ink. The tag would have a radar signature proper to it containing its ID. This would yield a flexible reading mode with a unit cost similar to that of a barcode. In concrete terms, due to its very low cost, this type of tag could be a disposable subway ticket with a reading mode similar to that of a high frequency (HF) RFID card, for example; but much more fluid. For all of these reasons, chipless RFID is an extremely promising technology, one that would open the way to the development of communication systems in which tags would have characteristics comparable to barcodes in terms of cost and simplicity of application. Last but not least, it would be an RF reading technology – wireless, and requiring no human interaction between the reader and the tag in order to recover the ID.

Index

D, E, F

G, H, I

Other titles from

iSTE

in

Networks and Telecommunications

2014

BATTU Daniel
New Telecom Networks: Enterprises and Security

BITAM Salim, MELLOUK Abdelhamid
Bio-inspired Routing Protocols for Vehicular Ad-Hoc Networks

CAMPISTA Miguel Elias Mitre, RUBINSTEIN Marcelo Gonçalves
Advanced Routing Protocols for Wireless Networks

CHETTO Maryline
Real-time Systems Scheduling 1: Fundamentals
Real-time Systems Scheduling 2: Focuses

EXPOSITO Ernesto, DIOP Codé
Smart SOA Platforms in Cloud Computing Architectures

MELLOUK Abdelhamid, CUADRA-SANCHEZ Antonio
Quality of Experience Engineering for Customer Added Value Services

OTEAFY Sharief M.A., HASSANEIN Hossam S.
Dynamic Wireless Sensor Networks

PEREZ André
Network Security

REMY Jean-Gabriel, LETAMENDIA Charlotte
LTE Standards
LTE Services

TANWIR Savera, PERROS Harry
VBR Video Traffic Models

VAN METER Rodney
Quantum Networking

XIONG Kaiqi
Resource Optimization and Security for Cloud Services

2013

ASSING Dominique, CALÉ Stéphane
Mobile Access Safety: Beyond BYOD

BEN MAHMOUD Mohamed Slim, LARRIEU Nicolas, PIROVANO Alain
Risk Propagation Assessment for Network Security: Application to Airport Communication Network Design

BEYLOT André-Luc, LABIOD Houda
Vehicular Networks: Models and Algorithms

BRITO Gabriel M., VELLOSO Pedro Braconnot, MORAES Igor M.
Information-Centric Networks: A New Paradigm for the Internet

BERTIN Emmanuel, CRESPI Noël
Architecture and Governance for Communication Services

DEUFF Dominique, COSQUER Mathilde
User-Centered Agile Method

DUARTE Otto Carlos, PUJOLLE Guy
Virtual Networks: Pluralistic Approach for the Next Generation of Internet

FOWLER Scott A., MELLOUK Abdelhamid, YAMADA Naomi
LTE-Advanced DRX Mechanism for Power Saving

JOBERT Sébastien *et al.*
Synchronous Ethernet and IEEE 1588 in Telecoms: Next Generation Synchronization Networks

MELLOUK Abdelhamid, HOCEINI Said, TRAN Hai Anh
Quality-of-Experience for Multimedia: Application to Content Delivery Network Architecture

NAIT-SIDI-MOH Ahmed, BAKHOUYA Mohamed, GABER Jaafar, WACK Maxime
Geopositioning and Mobility

PEREZ André
Voice over LTE: EPS and IMS Networks

2012

AL AGHA Khaldoun
Network Coding

BOUCHET Olivier
Wireless Optical Communications

DECREUSEFOND Laurent, MOYAL Pascal
Stochastic Modeling and Analysis of Telecoms Networks

DUFOUR Jean-Yves
Intelligent Video Surveillance Systems

EXPOSITO Ernesto
Advanced Transport Protocols: Designing the Next Generation

JUMIRA Oswald, ZEADALLY Sherali
Energy Efficiency in Wireless Networks

KRIEF Francine
Green Networking

PEREZ André
Mobile Networks Architecture

2011

BONALD Thomas, FEUILLET Mathieu
Network Performance Analysis

CARBOU Romain, DIAZ Michel, EXPOSITO Ernesto, ROMAN Rodrigo
Digital Home Networking

CHABANNE Hervé, URIEN Pascal, SUSINI Jean-Ferdinand
RFID and the Internet of Things

GARDUNO David, DIAZ Michel
Communicating Systems with UML 2: Modeling and Analysis of Network Protocols

LAHEURTE Jean-Marc
Compact Antennas for Wireless Communications and Terminals: Theory and Design

RÉMY Jean-Gabriel, LETAMENDIA Charlotte
Home Area Networks and IPTV

PALICOT Jacques
Radio Engineering: From Software Radio to Cognitive Radio

PEREZ André
IP, Ethernet and MPLS Networks: Resource and Fault Management

TOUTAIN Laurent, MINABURO Ana
Local Networks and the Internet: From Protocols to Interconnection

2010

CHAOUCHI Hakima
The Internet of Things

FRIKHA Mounir
Ad Hoc Networks: Routing, QoS and Optimization

KRIEF Francine
Communicating Embedded Systems / Network Applications

2009

CHAOUCHI Hakima, MAKNAVICIUS Maryline
Wireless and Mobile Network Security

VIVIER Emmanuelle
Radio Resources Management in WiMAX

2008

CHADUC Jean-Marc, POGOREL Gérard
The Radio Spectrum

GAÏTI Dominique
Autonomic Networks

LABIOD Houda
Wireless Ad Hoc and Sensor Networks

LECOY Pierre
Fiber-optic Communications

MELLOUK Abdelhamid
End-to-End Quality of Service Engineering in Next Generation Heterogeneous Networks

PAGANI Pascal *et al.*
Ultra-wideband Radio Propagation Channel

2007

BENSLIMANE Abderrahim
Multimedia Multicast on the Internet

PUJOLLE Guy
Management, Control and Evolution of IP Networks

SANCHEZ Javier, THIOUNE Mamadou
UMTS

VIVIER Guillaume
Reconfigurable Mobile Radio Systems

Printed and bound by CPI Group (UK) Ltd, Croydon, CR0 4YY
16/09/2026
14963022-0004